Reliable and Sustainable Electric Power and Energy Systems Management

Series Editors
Roy Billinton
Ajit Kumar Verma
Rajesh Karki

For further volumes:
http://www.springer.com/series/10855

Rajesh Karki • Roy Billinton • Ajit Kumar Verma
Editors

Reliability Modeling and Analysis of Smart Power Systems

Editors
Rajesh Karki
Department of Electrical and
Computer Engineering
University of Saskatchewan
Saskatoon, SK
Canada

Ajit Kumar Verma
Stord/Haugesund University College
Haugesund
Norway

Roy Billinton
Department of Electrical and Computer
Engineering
University of Saskatchewan
Saskatoon, SK
Canada

ISBN 978-81-322-1797-8 ISBN 978-81-322-1798-5 (eBook)
DOI 10.1007/978-81-322-1798-5
Springer New Delhi Heidelberg New York Dordrecht London

Library of Congress Control Number: 2014934187

Printed on acid-free paper

Springer is part of Springer Science+Business Media (www.springer.com)

Preface

This book series titled "Reliable and Sustainable Electric Power and Energy Systems Management" intends to provide a platform for researchers, planners, and policy makers to share their research outputs, ideas, and opinions on the critical issue of sustainable and reliable power and energy systems, and provide impetus for critical research in this highly important area for modern society in the context of a meshed and complex environment that is affected by events taking place throughout the world. This is the second volume of the book series, and is titled "Reliability Modeling and Analysis of Smart Power Systems."

Power systems throughout the world are undergoing significant changes creating new challenges to system planning and operation in order to provide reliable and efficient use of electrical energy. The appropriate use of smart grid technologies is an important drive in mitigating these problems, and requires considerable research activities. This book is focused on new innovative research from academia and industry on understanding, modeling, and quantifying the risks associated with different ways of implementing smart grid technologies in power systems in order to plan, operate, and maintain a modern power system economically and with an acceptable level of reliability. These are important issues in modern electric power systems anticipating the use of smart grid technologies.

This book consists of 13 chapters. Out of the 13 chapters in this volume, 10 are extended versions of papers presented at the PMAPS-2012 Conference, June 10–14, 2012 in Istanbul, Turkey. The biennial "Probabilistic Methods Applied to Power Systems" conferences are highly focused gatherings of international experts. Reliability analysis and evaluation of smart grid systems have been a major presentation and discussion topic in recent years and this activity is expected to continue in the future. The authors of all chapters in this book are actively involved in PMAPS and many of them participated in the recent conference in Istanbul.

The modernization of power systems with the application of smart-grid technologies is perceived to be a means to achieve reliability and efficiency with environmental compliance. Chapter 1 presents discussions on challenges in the ensuing reliability studies in the wake of an anticipated influx of smart-grid technologies. Though there are several shared visions among research organizations, industry, and academia on the philosophy of smart-grids, the means advocated to go about achieving them are quite disparate. The main objective of this chapter is a preparatory exercise to get ready for the anticipated challenges from the perspective of a reliability engineer.

There is a growing trend of integrating intermittent distributed generation into active distribution networks, which results in increased risks in the security of power supply to consumers. Plug-in Hybrid Electric Vehicles (PHEV) can be used to mitigate the intermittency of generated power, and to provide strategic supports to microgrids. Chapter 2 proposes a methodology to mobilize strategic microgrids for improved security of supply in an active distribution network considering PHEV. The proposed method incorporates many uncertainties that can be embedded in an active distribution network using a Monte Carlo simulation approach to quantify the feasible improvement in security of supply in PHEV based strategic microgrids.

A microgrid is vulnerable when operating in an islanded mode as the operators rely on local resources to ensure the balance between generation and load. The microgrid is then more sensitive to power quality issues such as voltage unbalance, caused by the connection of single-phase loads and sources. Electric vehicles can be envisaged as an active and flexible resource that can provide additional load or storage capacity to improve the microgrid emergency operating conditions. Chapter 3 analyzes the microgrid architecture and evaluates the impact of the active participation of electric vehicles on the microgrid frequency regulation in emergency conditions in an islanded mode.

Chapter 4 proposes an intelligent adaptive protection scheme for distribution systems integrated with distributed generators. The scheme utilizes digital directional over-current relays connected with a communication network to a central relaying unit. The presented scheme can identify the faulty section of the feeder and update the settings of the primary and backup relays to speed up the fault clearance. A linear optimization technique is used to coordinate the over-current relays whenever a change in the system topology is detected.

Application of intelligent electronic devices together with advanced peer-to-peer communication systems are recent developments in nonconventional protection systems in a smart grid platform. Failures of protection system components play a significant role in the outage events of breaker oriented power systems and protected components. Chapter 5 presents event tree analysis to evaluate the reliability of different alternative digitized protection schemes supported by an IED and a communication system. The results from the studies show that high penetration of advanced technologies in protection systems must also be associated with highly reliable components for enhanced system performance.

PHEV have received increasing attention in power system planning and operation. Uncontrolled charging and discharging of PHEV impose challenges in reliable system operation. They can however, support the system with ancillary services through “smart charging.” Chapter 6 introduces an approach to derive day-ahead charging profiles that minimize generation costs while respecting network and drivers’ end-use constraints, as well as taking into account the uncertainty in driving patterns. A probabilistic method is used to model individual driving behavior, which is then integrated as virtual batteries in the power system evaluation model.

Chapter 7 proposes a copula-based stochastic approach to derive the load demand of a fleet of domestic PHEV. Home arrival time, daily travelled distance, home departure time, driving habits, and road traffic conditions are the most effective factors on the load demand of PHEV. Deterministic methods used previously to model these factors cannot recognize the stochastic nature of these factors. A probabilistic approach is used in this chapter to create appropriate probability density functions from data collected from randomly selected vehicles. Monte Carlo simulation is used to generate random samples and estimate the probability density function of the hourly aggregated load of the PHEV. The power delivered to the PHEV can then be obtained and used in analyses, such as network planning, load management, and probabilistic load flow, as well as sitting and sizing problems.

As mentioned above, a PHEV consumer exhibits a stochastic behavior. Probabilistic methods can be used to effectively incorporate these inherent uncertainties in real world applications. Chapter 8 incorporates PHEV uncertainties into home load controlling to achieve economic benefits. An optimization problem is formulated based on mixed-integer programming, and numerical studies are conducted in order to illustrate the effectiveness of the developed model. The presented method can be used to economically schedule the consumption periods of responsive appliances as well as the PHEV battery charge/discharge periods. The results show that the vehicle to grid capability of PHEV can dramatically reduce energy costs.

Smart grid technologies, such as AMI, smart monitors, smart appliances, and smart controllers, will facilitate load management activities. Chapter 9 reviews the effects of advanced load management on smart grid reliability. The existing reliability indices may require modifications and additions in the new environment due to significant changes in load management methods in future smart grids. This chapter discusses the changes in reliability indices of distribution systems, and proposes some modifications for these indices.

Chapter 10 investigates the effect of reliability worth in the optimal economic operation of a small autonomous power system (SAPS) that is based on renewable energy sources (RES) technologies. The optimization procedure is implemented with a combined genetic algorithm (GA) and local search procedure. This chapter also examines the effects of component forced outage rates and uncertainty in weather and cost data using Monte Carlo simulation.

Nonconventional and smart technology is receiving attention in different areas of power system planning, operation, and maintenance. Chapter 11 describes the use of real time data in the maintenance of offshore wind turbines. This chapter focuses on the role of condition monitoring to lower costs and risks associated with short-term reliability and long-term asset integrity. The studies carried out in this chapter illustrate how a monitoring system can be optimized for risk reduction and for the reduction of the expected structural integrity management costs.

The application of reliability centered maintenance to a fleet of wind turbines is presented in Chap. 12. Maintenance records were analyzed to identify the key components and failure modes. The chapter provides discussions on corrective actions and implementation issues to mitigate such failures, and the need for a robust set of RCM tools to better quantify and minimize operational expenditures.

Chapter 13 presents an optimization model that incorporates diagnostics, economy, and reliability to determine cable segment replacement in distribution systems. The optimization is based on diagnostic measurements, which typically could be made online with temperature sensors and/or partial discharge detection. The method identifies cable segments for replacement under a budget constraint and encourages replacement of continuous segments of cable.

Smart grid technologies are receiving significant attention in the electric power industry and in academia. Their applications in power systems are still in the development stage, and are expected to encounter various challenges. This book provides valuable insight and discussions on development of methodologies to assess risk and reliability in smart grid applications in the planning, operation, and protection of power generation and distribution systems, and focuses attention on reliability related maintenance of system components.

Rajesh Karki
Roy Billinton
Ajit Kumar Verma

Contents

Contributors

Ahmed Saleh Alabdulwahab Electrical and Computer Engineering Department, College of Engineering, King Abdulaziz University, Jeddah, Saudi Arabia

Farrokh Aminifar School of Electrical and Computer Engineering, College of Engineering, University of Tehran, Tehran, Iran

Göran Andersson Power Systems Laboratory, ETH, Zurich, Switzerland

Graham W. Ault Institute for Energy & Environment, University of Strathclyde, Glasgow, UK

Roy Billinton University of Saskatchewan, Saskatoon, SK, Canada

Akbar Ebrahimi Department of Electrical and Computer Engineering, Isfahan University of Technology, Isfahan, Iran

Mahmud Fotuhi-Firuzabad Center of Excellence in Power System Management and Control, Department of Electrical Engineering, Sharif University of Technology, Tehran, Iran

Pavlos S. Georgilakis School of Electrical and Computer Engineering, National Technical University of Athens (NTUA), Athens, Greece

Masoud Aliakbar Golkar K. N. Toosi University of Technology, Tehran, Iran

Clara Sofia Gouveia INESC TEC—INESC Technology and Science (formerly INESC Porto) Porto, Campus da FEUP, Porto, Portugal

Mohamed S. Hassan Electrical Engineering, American University of Sharjah, Sharjah, UAE

Patrik Hilber Electromagnetic Engineering, KTH, Royal Institute of Technology, Stockholm, Sweden

Syed Islam Department of Electrical and Computer Engineering, Curtin University, Perth, Australia

Dilan Jayaweera Department of Electrical and Computer Engineering, Curtin University, Perth, Australia

Rajesh Karki University of Saskatchewan, Saskatoon, Canada

Yiannis A. Katsigiannis Department of Natural Resources & Environment, Technological Educational Institute of Crete, Chania, Crete, Greece

Gerd H. Kjølle SINTEF Energy Research, Trondheim, Norway

João Peças Lopes INESC TEC—INESC Technology and Science (formerly INESC Porto) Porto, Campus da FEUP, Porto, Portugal

David McMillan Institute for Energy & Environment, University of Strathclyde, Glasgow, UK

Carlos L. Moreira INESC TEC—INESC Technology and Science (formerly INESC Porto) Porto, Campus da FEUP, Porto, Portugal

Marios N. Moschakis Department of Electrical Engineering, Technological Educational Institute of Larissa, Larissa, Greece

Amir Moshari Department of Electrical and Computer Engineering, Isfahan University of Technology, Isfahan, Iran

Ahmed H. Osman Electrical Engineering, American University of Sharjah, Sharjah, UAE

Ehsan Pashajavid K. N. Toosi University of Technology, Tehran, Iran

Mohammad Rastegar Center of Excellence in Power System Management and Control, Department of Electrical Engineering, Sharif University of Technology, Tehran, Iran

Paulo Ribeiro INESC TEC—INESC Technology and Science (formerly INESC Porto) Porto, Campus da FEUP, Porto, Portugal

Amir Safdarian Center of Excellence in Power System Management and Control, Department of Electrical Engineering, Sharif University of Technology, Tehran, Iran

Kjell Sand SINTEF Energy Research, Trondheim, Norway

Mohamad Sulaiman Electrical Engineering, American University of Sharjah, Sharjah, UAE

Sebastian Thöns Division 7.2 Buildings and Structures, BAM Federal Institute for Materials Research, Berlin, Germany

Vijay Venu Vadlamudi Norwegian University of Science and Technology, Trondheim, Norway

Marina González Vayá Power Systems Laboratory, ETH, Zurich, Switzerland

Bio-sketch of Editors

Dr. Rajesh Karki is Associate Professor at the University of Saskatchewan, Canada. During 2005–2012, he chaired the Power Systems Research Group at the University of Saskatchewan. He has earlier worked for academic institutions and different industries in Nepal, and for GE Industrial Systems, Peterborough, ON, Canada. He has a B.E. degree in electrical engineering from the Regional Engineering College (renamed National Institute of Technology), Durgapur, West Bengal, India, and Masters and PhD degrees in electrical power engineering from the University of Saskatchewan, Canada. He has served in various capacities in conferences, workshops, and guest lectures in Canada and abroad and has published over 80 papers in reputable international journals and peer reviewed conferences. Dr. Karki has completed several consulting projects on system planning and reliability for Canadian electric utilities. He is a senior member of the IEEE, and Professional Engineer in the Province of Saskatchewan, Canada. His research interests include power system reliability and planning, and reliability modeling and analysis of renewable energy systems.

Dr. Roy Billinton is Professor Emeritus in the Department of Electrical Engineering, University of Saskatchewan, Saskatoon, Canada. He obtained his Bachelor's and Master's degrees from the University of Manitoba, Winnipeg, MB, Canada, and PhD and D.Sc. degrees in electrical engineering from the University of Saskatchewan, Saskatoon, Canada. He was with the System Planning and Production Divisions, Manitoba Hydro, Canada. In 1964, he joined the University of Saskatchewan, where he served as the head of the Electrical Engineering Department, Associate Dean for Graduate Studies, Research and Extension, and the Acting Dean of Engineering. He has authored or coauthored 8 books and more than 940 papers on power system reliability evaluation, economic system operation, and power system analysis. Dr. Billinton is Fellow of the IEEE, Engineering Institute of Canada, Canadian Academy of Engineering, Royal Society of Canada. He is also Foreign Associate of the United States National Academy of Engineering, and Professional Engineer in the Province of Saskatchewan, Canada. In 2010, the IEEE-PES honoured Dr. Billinton by initiating the Roy Billinton Power System Reliability Award.

Dr. Ajit Kumar Verma is Professor in Engineering, Stord/Haugesund University College, Haugesund, Norway. He is also Guest Professor at Lulea University of Technology, Sweden. He was earlier Professor, Department of Electrical Engineering at IIT Bombay, Mumbai, India. Prof. Verma is Chairman of recently constituted Special Interest Group on System Assurance Engineering and Management of Berkeley Initiative in Soft Computing, UC Berkeley, USA. He is author of books titled *Fuzzy Reliability Engineering-Concepts and Applications* (Narosa), *Reliability and Safety Engineering* (Springer), *Dependability of Networked Computer Based Systems* (Springer) and *Optimal Maintenance of Large Engineering System* (Narosa) and has over 225 publications in various journals (over 100 papers) and conferences. He has served as Editor-in-Chief of *OPSEARCH* as well as Founder Editor-in-Chief of *International Journal of System Assurance Engineering and Management* both journals published by Springer, and the Editor-in-Chief of *Journal of Life Cycle Reliability and Safety Engineering*. He has been a guest editor of a dozen issues of international journals including IEEE *Transactions on Reliability*.

Reliability-Centric Studies in Smart Grids: Adequacy and Vulnerability Considerations

Vijay Venu Vadlamudi, Rajesh Karki, Gerd H. Kjølle and Kjell Sand

1 Introduction

Of late, there has been a surge in embracing the various smart grid (SG) visions in the power sector. There is a burgeoning literature on SGs for various blue prints to realize them. Utilities world-wide have begun to initiate piece meal programs to selectively implement features that aim to improve the "smartness" quotient of the existing grids. Most of the focus on SGs entails better collection and utilization of system information in the various decision making processes in operational and planning horizons. The primary motivation in making a power system SG compliant is to have viable alternatives that address the concerns for explicit and imminent infrastructural expansion to match the demand for growing energy needs, all while going green. This transition is invariably reliant on the extensive use of information and communication technologies (ICT) to add the dimension of intelligence to various functionalities of a power system, which "moves it forward."

This chapter presents some thoughts on challenges in the ensuing reliability studies in the wake of perceived influx of SG technologies. Though there are several shared visions among research organizations, industry, and academia on the philosophy of SGs, the means advocated to go about achieving them are quite disparate. It must be pointed out that the goal of this chapter is not a discussion on how to realize the vision of SGs, but a preparatory exercise on readying for the anticipated

V. V. Vadlamudi (✉)
Norwegian University of Science and Technology, Trondheim, Norway
e-mail: Vijay.Vadlamudi@ntnu.no

R. Karki
University of Saskatchewan, Saskatoon, Canada
e-mail: Rajesh.Karki@usask.ca

G. H. Kjølle · K. Sand
SINTEF Energy Research, Trondheim, Norway
e-mail: Gerd.Kjolle@sintef.no

K. Sand
e-mail: Kjell.Sand@sintef.no

R. Karki et al. (eds.), *Reliability Modeling and Analysis of Smart Power Systems*, Reliable and Sustainable Electric Power and Energy Systems Management, DOI 10.1007/978-81-322-1798-5_1, © Springer India 2014

challenges from the perspective of a reliability engineer. A definition of SG, which could bring into focus the relevance of the task undertaken, as put forward by the Electric Power Research Institute (EPRI) [1] is as follows:

> A Smart Grid is one that incorporates information and communications technology into every aspect of electricity generation, delivery and consumption in order to:
>
> - Minimize environmental impact
> - Enhance markets,
> - Improve reliability and service, and
> - Reduce costs and improve efficiency.

It helps to have a systems view to successfully emerge from the diversity of thoughts on realizing SGs. The National Energy Technology Laboratory (NETL) [2] presents one such view, where the eventual emphasis is on the development of metrics that gauge whether the performance required of the grid as dictated by its desired characteristics is achieved through the deployment of relevant technologies. These metrics in turn help in identifying the extent of necessary fine tuning of the realistic targets aspired through the stated performance requirements in the iterative procedure of achieving process maturity. For example, for each of the four goals mentioned above by EPRI in its definition of SGs, appropriate metrics could be endeavored that aptly quantify the justification of SG benefits. Some of these might turn out to be competing metrics, in which case multi-criteria decision making must be resorted to for an overall assessment of the cumulative benefits.

Almost every working definition on SGs emphasizes on the term "reliability," more often than not in a qualitative sense. From a power system reliability engineer's perspective, this quantification can form a tangible basis for both investment planning and operational planning with respect to reliability worth. While making qualified guesses about the initial steps towards possible quantification of reliability benefits of relevant SG technologies at various power system hierarchical levels (HLs) [21], this chapter also includes notes on the relevance of cross-sector risk and vulnerability analyses that are vital to the study of interdependencies between different infrastructures. The preliminary analysis concerns the potential improvement in the adequacy domain of reliability status of smart grid-oriented power systems (SGOPS)[1], which in turn contributes to steady-state security of supply. The premise of the chapter has its scope limited only to the prevailing static conditions that do not include system dynamic and transient disturbances, i.e., steady-state security constrained adequacy assessment.

[1] While "SG" is used in this chapter to associate a vision/philosophy, "SGOPS" is used to refer to a power system embarking (or about to embark) on realizing the vision(s).

2 Reliability Studies and Allied Challenges in SGOPS

Reference [3] provides an overview of the challenges of SG reliability objectives and identified the reliability impact of four major SG resource types—renewable resources, load management/demand response, storage devices, and electric transportation. They observed how a flattening of load profile through an optimal mix of the SG resources, though initially tending to improve reliability by lowering the peak, would eventually intensify the reliability challenges as the flattened load grows. Their work included the proposal of an evolutionary IT infrastructure architecture for SGs to cope with such challenges. Reference [4] emphasizes on the heavy dependence of SG reliability on information architecture and highlighted the need for distributed database structure that provides the basis for real-time information architecture and substation information architecture. The North American Electric Reliability Corporation (NERC) recently presented a preliminary report [5], expanding on the reliability considerations arising from the large scale integration of SG technologies and devices, concluding that it can indeed improve bulk power system reliability if and only if appropriate tools and models are developed to cope with newer planning and operational strategies. All these studies point to an imminent need for increase in reliability-related research and development that captures the impact of ICT functionalities into power systems.

The International Energy Agency (IEA), in its technology roadmap for SGs [6], lists a variety of technologies that span the entire electricity system at various functional levels of generation, transmission, and distribution, and further delineates the constituent hardware, and systems and software requirements. From a reliability perspective, to note the explicit improvement in the adequacy status of an SGOPS, identification of resource adequacy contributors from its perceived architectural composition is the key beginning step in attempts at quantifying the purported reliability impacts.

This was initiated in an earlier paper by the authors [7], where ICT was projected as a networked feedback control system, whose permeation in the power system now modifies the architectural composition of existing grids with the inclusion of resources related to technologies such as demand response (DR), distributed generation (DG), sensors, wide area measurement systems (WAMS), advanced metering infrastructure (AMI), advanced power electronics (APE), distributed automation (DA), and integrated communications (IC), to name a few. Some thoughts on the basic evolution of metrics that quantify the accompanying reliability benefits are presented in the next section.

IEA [6] also points out how the long-term system adequacy concerns, especially the ones introduced by the deployment of variable generation technology, could be additionally addressed through "flexibility" mechanisms in the operational horizon such as interconnection with adjacent markets, citing the example of the Nordic electricity system. Perhaps, latest studies such as the ones on balance market integration [8] could also aid in eventually studying the reliability contributions of such flexibility mechanisms that shall have percolated to the surface. Though such mechanisms currently seem far removed from the definitive agenda of reliability improvement alone, seen from the above-mentioned IEA perspective, their prominence in indirectly harnessing the potential of renewable energy sources can be hardly overstated.

In order to improve resource adequacy in SGOPS, there must be strategic deployment of power system technology improvisations, usually made possible by ICT arrangements [7]. By acting as a feedback control system, ICT ensures that technologies do not play next to each other, but with each other. It must be highlighted that technologies enable "functions." They can eliminate failures or themselves be sources of failures. Hence, two components of reliability must be considered for the study of SGOPS:

1. Stand-alone reliability of the technological constituent/infrastructure, and
2. Contribution of the technological constituent/infrastructure to the overall functional reliability.

Modular studies in the former category have been recently reported [9–13], and are gradually on the rise. However, it is in the latter category that very little or no research has been reported, partly because of the difficulties encountered in explicit delineation of the many functional capabilities of ICT. In this context, we opine that one possible way of addressing this would be to rely on the multi-layered approach (encompassing physical, communication, functional, and business layers) of use-cases.

Ideally speaking, to get a definite picture of functional reliability benefits, post-facto analysis of sufficient sample data/information from the use-cases through hypothesis testing is the best recourse. The testable hypothesis could be that ICT will improve system fault/failure diagnosis and prognosis, thus correlating to a "reliability" function. This is just a beginning step in capturing the reliability performance through sample historical collection and analysis of system data. However, to estimate the future performance of designated configurations, analysis based on additional newly developed theoretical models is necessary for reliability prediction.

EPRI maintains a use-case repository on its website [14] based on the reported experiences of SG pilot initiatives across the industry. In addition to the routine performance analysis studies, such use-cases could be employed to embark on functional reliability-related studies. The European Commission Directorate-General for Energy has recently presented a standardization mandate [15] to European Standardization organizations to support European SG deployment, where the need for a hierarchical collection and harmonization of use-cases is duly stressed upon. The proliferation of use-case based approaches can be viewed as advantageous for a reliability point of view as well. In general, other key aspects to be considered in ICT reliability evaluation are information integrity, communication failures due to data traffic or other reasons, and hardware and software fault tolerance.

With the expected extensive permeation of ICT in SGOPS, there emerges a critical interdependent infrastructure, tightly coupled to the existing power system infrastructure, mandating a "system of systems" perspective [16] to deal with the emergent behavior of these complex adaptive systems. Though component additions increase the availability, the risk of failure scenarios due to complex interactions increases. In this new environment, the risk of extraordinary events such as catastrophic cascade of failures substantially increases. As pointed out in [17], interdependent networks are extremely sensitive to random failures, and failure of even a small fraction of nodes in one network can propagate iterative failures in the others. Further, risk analysis of interdependencies in critical infrastructures is necessary for identifying vulnerabilities, ensuring emergency preparedness and prioritizing risk reducing measures [18]. These issues are all to be viewed as allied reliability-related challenges.

More specifically, interdependency is defined as "a bilateral relationship between two infrastructures through which the state of each infrastructure influences or is correlated to the state of the other" [16]. A taxonomy comprising six different dimensions for describing infrastructure interdependencies is proposed in [16], which can also be potentially employed in studying interdependent failures leading to exacerbating conditions in SGOPS. Reference [19] points out how robustness of power system is dominant, but vulnerability latent. They explain that SGOPS "is robust enough to operate but has potential vulnerability" which might lead it into cascaded failure modes. Thus, risk and vulnerability studies are equally vital to ensuring security of supply, complementing the goals of adequacy assessment.

Amidst all the automation expected to be brought on by ICT, human performance reliability studies assume more significance than ever before. Though design and technology are pivotal to the reliability enhancement, the role of operators in ensuring operational reliability can be hardly overlooked. Once again, studies on the human reliability component in a man–machine system [20] have a direct bearing in the evolving SGOPS.

3 Reliability Benefits of Relevant SG Technologies at Hierarchical Levels

Fundamentally, quantification of load point adequacy (as seen by the respective functional zones of power system) or system adequacy is done by way of evaluating the frequency and the duration associated with various failure events (though practically, at HLI only the loss of load probability (LOLP) is of relevance). Probability of occurrences of failures can be derived by multiplying the corresponding frequency and duration values of events. Since these basic indices do not reflect the severity of events, additional indices such as expected energy/demand not served are computed at HLI and HLII. In order to capture the localized impact of failure events at the distribution level (DL), customer-oriented indices such as SAIFI (system average interruption frequency index), SAIDI (system average interruption duration index), CAIDI (customer average interruption duration index), CAIFI (customer average interruption frequency index), and ASAI (average service availability index) are also in vogue [21]. Energy that could not be delivered to customers due to interruptions, also a measure of lost revenue for the utility, is quantified by energy not supplied (ENS) and average energy not supplied (AENS). The role of various SG

Table 1 Specific failure events of interest in adequacy assessment at various hierarchical levels

Hierarchical level	Failure events
HLI	Load loss
HLII	Load loss, network condition—related events such as line overloads and bus voltage violations
HLIII (distribution zone)	Customer interruptions (momentary and sustained)

technologies in reliability improvement could be noted by gauging the observed/estimated values of these indices. Specific failure events of interest in adequacy assessment at various HLs are summarized in Table 1.

3.1 SG Technology at HLI and HLII

Load loss at HLI could be due to increased demand coupled with disproportionate generation, or because of random failures in the generating units, or a combination of these factors. Load loss at HLII, if in spite of adequate generation, could be a result of operational congestion on the transmission lines or random failures in the transmission lines. DR, distributed energy resources (DER) and APE technologies, mitigate the failure events at HLI and HLII. DR and DER technologies at HLI and HLII are aimed at lowering the frequencies of failure events by mainly addressing the capacity deficiency issues.

Though sounding like a backhanded approach, energy conservation (made possible through DR activities and AMI) quite effectively actuates the adequacy improvement. DR programs yield fast responding operating reserve in terms of interruptible load. Flattening of load curves through DR results in improved load factors. DSM emphasizes on incentivization to change the consumption modes of customers. Encouraging newer energy usage patterns in line with available generation alleviates the stress on resource mobilization to avoid capacity deficiency. Development of ICT that supports DSM is a key driver to the adoption of DR programs in the evolving regime. Quantifying the expected impacts of peak clipping, load shifting, valley filling, and energy conservation schemes of DSM on load curves at all the load buses in HLII can enable studies on composite system reliability indices [22]. Such studies are once again expected to gather momentum given the ICT permeation which provides sufficient infrastructural capabilities for the extensive realization of DSM.

DER increases the diversity of resource-mix options, also contributing to emission reductions. APE results in the minimization of real power load curtailments (a direct consequence of minimizing line congestions through modified allowable line thermal rating settings) at HLII. Phasor measurement units (PMU) and WAMS contribute to reduced frequency of failure events at HLII on account of the wide area situational DL awareness made accessible to system operators in detecting anomalies in network conditions, readying them for anticipated remedial actions. How the reduction in frequencies and durations of failure events due to SG technology further translates to improvement in the bulk power system reliability indices [21] must be investigated for analyzing the more tangible reliability benefits of adopting SG technologies.

3.2 SG Technology at the Distribution Level

Supervisory control of distribution systems is possible through the distribution management systems (DMS) component of SG technology. By using the decision support tools of this facility (outage management systems), in terms of analysis, interruption statistics could be effectively compiled for a retroactive examination of distribution reliability indices. In terms of operational effectiveness, duration of customer interruptions could be drastically reduced by means of faster fault location and service restoration, which are in turn enabled by the deployment of sensors and advanced metering infrastructure. This has a direct impact on reduction in SAIDI and CAIDI. Reliability centered maintenance (RCM), which taps into the database of maintenance data archives based on expert systems, due to SG technology at the DL is expected to gain relevance. Alongside distribution automation (which provides real-time information on the statuses of voltages, feeders, protection equipment, etc.), RCM has the potential to reduce sustained interruptions before they turn into permanent outages through continuous condition monitoring, thus lowering SAIFI. The eventual benefit of SG technologies at the DL is the lowering of the ENS index and the customer interruption costs due to subsequent reduction in the frequency and the duration of interruptions. For the foreseeable future, the radial nature of power systems is expected to be preserved. Methods such as RELRAD [23] continue to retain their effectiveness. Active distribution systems necessarily do not mean that the system will be made meshed. However, inclusion of active functions in the reliability models at the DL is increasingly desirable.

Several direct and indirect effects of DG on system reliability at the DL have been pointed out in the United States Department of Energy Report [24]. It has been highlighted how a distributed network of smaller sources is responsible for a greater level of adequacy in comparison with a centralized system with few larger sources, leading to reduction in the magnitude and the duration of failures. “DG could be used directly to support local voltage levels and avoid an outage that would have otherwise have occurred due to excessive voltage sag” [24]. DG can contribute to grid connected power demand reduction, which imposes less stresses on substation equipment and grid components, resulting in the indirect benefit of decreased component failure rates. Microgrids, provided they are “integrated” into the system and not “tolerated,” will contribute to increased capacity credits at HLI [25].

3.3 Value-Based Reliability Investments in SG Technology

Value-based reliability planning has been one of the guiding principles for power system reinforcements. "A value-based planning method is appealing in that it attempts to mathematically compare the cost of unreliability with the cost of providing additional reliability in order to determine whether the system enhancements are justified" [26]. With load loss and the associated energy loss, the situation of related customer outage costs arises, which is what ultimately constrains the reliability cost-effectiveness of SG investments. Investment costs of bringing about reliability improvements to a system must not exceed the value of consequential economic losses suffered by the end users if such improvements were not in place.

For the realization of SG innovations that foster reliability at all hierarchical levels, keeping this philosophy in mind, comprehensive studies on customer outage costs must be carried out. Tangible "inconvenience indices" must be floated that can translate into monetary equivalents that serve as basis for the feasible extent of SG technology deployment. This can be achieved when there is accessibility to reasonable estimates of customer interruption costs in the emergent scenario. The variation in the various customer damage functions (CDF) [21], influenced by the consumption-behavior changes driven by demand response technologies and the changes to the customer interruption costs, with respect to the reduction in frequency and duration of failures consequent to the adoption of SG technology can in fact be evaluated. In relation to investigating how the costs of providing an acceptable quality should be balanced against the value of quality, [27] presents the methodology and survey results of the most recent Norwegian customer survey on consumer valuation of interruptions and voltage problems. Such comprehensive studies assume increasing significance in research on value-based reliability investments in SGOPS.

While at the DL, the quantum of reinforcements in consideration for such reliability trade-off studies is with respect to the installation of intelligent devices that form the backbone of ICT facilities, it mainly boils down to the provision of generation reserves at HLI and HLII. Value-at-risk (VaR) alongside well-being analysis (WBA)- based studies can let the customers have an explicit say that is economically beneficial in the reliability provisions guaranteed by the arrangement of generation reserves [28].

The distinct possibility of demand side management brought on by improved communication with customers through AMI (smart meters) can enable power system planners and operators with a variety of options. A sample situation in this regard is as demonstrated in Fig. 1. There could be unexpected failures or a pressing need for planned maintenance activities in the installed generation capacity, which reduce the established reserve margins from level A to level B in Fig. 1. Upon sending energy conservation signals to the customers through AMI, if significant peak load reduction cannot alter the shape of the load duration curve favorably, before taking expensive measures such as resorting to interconnected assistance or readying for costlier stand-by generation, value-based operational decisions could be made through quick online studies.

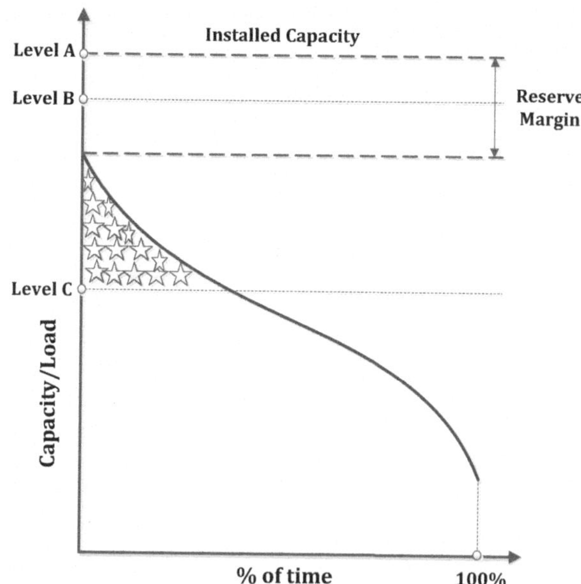

Fig. 1 Sample illustration of load duration curve for value-based reliability options (*starred portion* depicts energy lost)

Say, consequences of clipping the load duration curve (LDC) up to level C in Fig. 1 could be studied in relation to the arising outage costs vs. costs associated with hard limits on reserve provisions. Even though there might be a load loss (an indicator of unreliability), what matters is the customers' satisfaction with respect to monetary compensation, should the need arise. Thus, a meta-study of customer outage costs can prove beneficial in the planning and the operational horizons from a reliability-worth stand point.

A tabular representation of reliability benefits of relevant technologies at HLI, HLII and DL is shown in Table 2. It is expected that future research would be focused on improved quantification of reliability benefits of SG functionalities than what is permissible within the frameworks available so far.

4 Reliability-Centric SG Research in Norway

One of the most important areas of activity of the Norwegian Smartgrid Centre (NSgC) [29] is to initiate and co-ordinate research in SG technology in Norway. One of its flagship projects concerns the development of Next Generation Control Centres (NGCC), where studying the reliability impact of closer operational integration of electrical and ICT infrastructures is an explicitly stated goal. Additionally, use-case based approaches for performance assessment are being built-upon. It has been identified how the observations from Table 2 pertaining to the lowering of frequency and duration of failure events hold ground only for individual piecemeal implementation (bottom-up approach) of SG technologies from the view point of

Table 2 Impact of various SG technologies on frequency and duration of failure events at HLs

SG Technology	Hierarchical Level	Reliability Contribution		
		Frequency of Failure Events	Duration of Failure Events	Remarks
DR	HLI, HLII	↓		Addresses capacity deficiency; Improves load factors; Alters LDC; Creates non-spinning reserve.
DG	DL	↓	↓	Creates contingency reserves; Improves component failure rates of grid components; Avoids local voltage related outages.
DER	HLI, HLII	↓		Addresses capacity deficiency; Reduces emissions; Increases capacity credits.
APE	HLII	↓		FACTS devices and Dynamic Thermal Circuit Rating (DTCR) estimates alleviate congestion that might lead to load curtailment.
	DL		↓	Helps integrate DG into power systems; Enables rapid switching actions.
PMU, WAMS	HLII	↓		Situational awareness for remedial actions.
AMI	DL	↓	↓	Complements DR and DMS; Also helps in the detection of power quality issues and power loss.
DA	DL	↓	↓	Improves outage response and restoration strategies; Reconfigures network; Protection systems decrease cascading outages
DMS	DL		↓	Supervisory control helps in the outage management.
RCM	DL	↓		Enables continuous condition monitoring.
IC	HLI, HLII, DL	↓	↓	Complements every technological constituent of SG.

contribution to adequacy alone. Upon simultaneous deployment of various technologies, only a top-down approach will be able to reveal the impact of failures in the functional interplay of these technology constituents. Unless relevant mathematical models which capture the consequent interdependency effects between the electrical and ICT infrastructures through analytical contingency analysis simulation are in place, the cumulative impact on frequency and duration of failure events cannot be anticipated. In this connection, vulnerability analysis (briefed in the next section) is deemed to be of vital importance.

The focus in the NGCC project is centered around distribution level SG reinforcements as is usually the case with several pilot projects around the world. One of the reliability-centric research goals of NGCC is to uncover the correlation between interoperability and reliability. Interoperability concerns are primarily rooted in performance enhancement. The GridWise Architecture Council defines interoperability [30] as "the capability of systems or units to provide and receive services and information between each other, and to use the services and information exchanged to operate effectively together in predictable ways without significant user intervention." Both technical and informational interoperability dimensions must be ensured to successfully integrate SG devices and assets into the system. It has been qualitatively affirmed [30] that interoperability yields reliability benefits. The IEEE Std. 2030–2011 guidelines [31] recommend adoption of standards as a potential tool for enhancing interoperability. In line with this, research in the NGCC is intended to be directed towards the NIST priority areas for implementation of Standards [32] to quantify the success of interoperability from a reliability vantage point with standards as the basis.

A new interoperability layer can be introduced in the original SG hierarchical level classification framework for adequacy studies presented in [7], as shown in Fig. 2.

5 Vulnerability

The electric power system and the ICT system are both complex infrastructures. As mentioned before, this complexity increases in SG along with new technologies, components, and functional interdependencies (see, e.g., [16, 18] for descriptions of dependencies). While the reliability and robustness of SGOPS in general is expected to increase, the risk of rare events might increase due to vulnerabilities caused by dependencies, new components, cyber threats, new operating scenarios, etc. Such rare events which might lead to high impact in terms of interruption duration, energy not supplied, and societal costs, are generally regarded to have low probability of occurrence. Traditional probabilistic methods applied to power systems are often not well suited to capture rare events, for instance to identify potential cascading effects and the consequences [18]. Vulnerability analysis is needed to complement the reliability analysis, in particular to analyze interdependencies and the influence of human factors. For this purpose it is necessary to combine different qualitative and quantitative methods [33]. Various authors have proposed methods for vulnerability analysis emphasizing dependencies, and a few such studies are focused on power systems. A framework for vulnerability analysis of power systems is presented in [34, 35]. This framework is used in [18] together with contingency and reliability analyses, and the results are used as input to the risk analysis of interdependencies with other infrastructures such as ICT. Analysis of interdependencies may either focus on the causes, the consequences, or both. In the approach presented in [18], a cascade diagram is used to investigate the consequences. The cascade diagram gives an overview of interdependencies in a structured manner. It resembles an

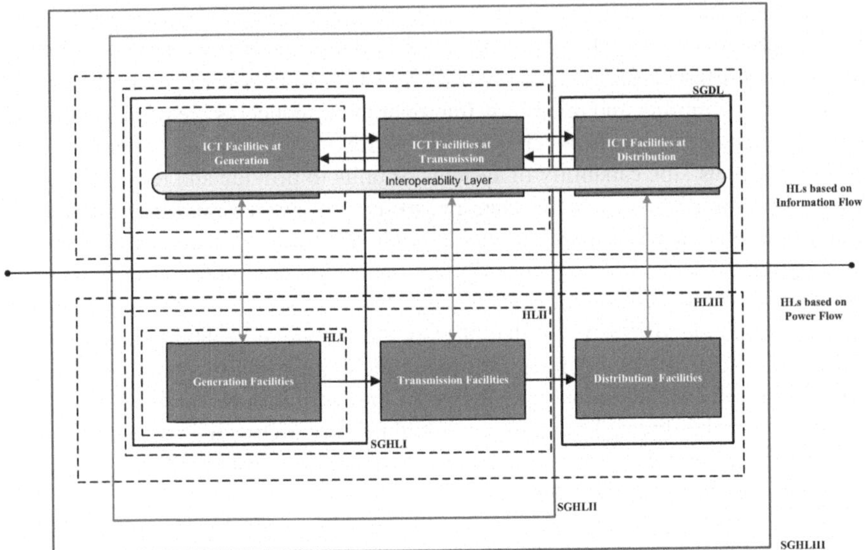

Fig. 2 Modified SG hierarchical level classification for adequacy studies. The nomenclature is as follows: smart grid distribution level (*SGDL*), smart grid hierarchical level I (*SGHLI*), smart grid hierarchical level II (*SGHLII*), smart grid hierarchical level III (*SGHLIII*). (Adapted from [7])

event tree but focuses on the chain of cascading events. This approach is a promising starting point for risk analysis of wide-area interruptions and the consequences for other infrastructures. It is necessary, however, to further develop these kinds of methods for the integrated power and ICT system represented by SG.

Some of the relevant information documented on monitoring vulnerability in power systems [35] is briefly presented here. Vulnerability is a characteristic of the system comprising susceptibility and coping capacity. While vulnerability is an internal characteristic of the system, risk can be defined as a combination of the probability and consequence of an undesired event. Vulnerability may affect both the probability and the consequence, and is as such a component of risk.

Fault statistics is probably the best available data basis for risk evaluation regarding causes of power system failures and their consequences in terms of interruptions to load points. Figure 3 shows examples of indicators in use today based on the fault statistics. Fault frequency describes the result of exposure to threats and the susceptibility towards these threats. Energy not supplied (ENS) adds information about the coping capacity, i.e., the consequence of the undesired event measured as interrupted load and duration. Expected interruption costs (EIC) add information about the societal consequences for different end-users.

Outage frequency, ENS and EIC are lagging indicators describing past performance. They give aggregate information about vulnerability. However, there is a need for indicators providing information about each of these dimensions: threats, susceptibility, coping capacity, and potential consequences. Obviously the above mentioned indicators are inadequate for the purpose of monitoring the various

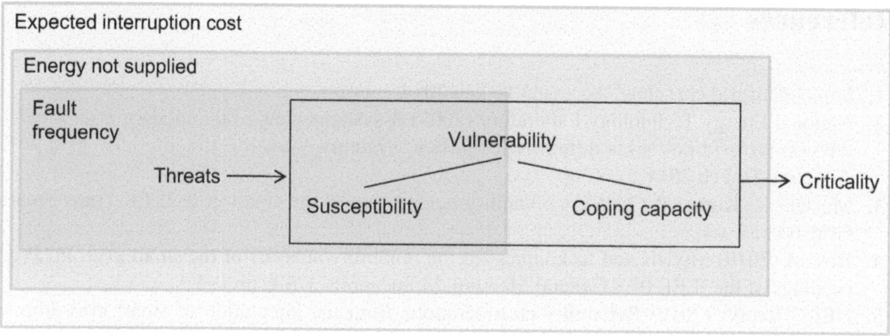

Fig. 3 Examples of indicators describing parts of vulnerability [35]

dimensions of vulnerability, since the effects are often aggregated. It is necessary to develop leading indicators capable of predicting the development of vulnerability to provide information about risk exposure related to extraordinary events in a changing power system. Thus, in light of extensive permeation of ICT into the existing power grid, vulnerability analysis of the interdependent infrastructures is indispensable to complement the goals of reliability studies in SGs.

6 Conclusions

An attempt has been made in this chapter to identify the basic metrics associated with SG technology deployment, in terms of frequency and duration of failure events. The discussion presented is an attempt at the initial steps towards the evolution of eventual comprehensive metrics that can underscore the reliability attributes of SGOPS. Anticipated reliability benefits of relevant SG technologies at various power system hierarchical levels have been outlined. Several challenges in SG studies from the reliability-standpoint (such as cross sector risk, vulnerability, and human performance reliability studies) have been pointed out. Especially, vulnerability analysis is considered to gain significant attention in light of the growing interdependencies between electrical and ICT infrastructures in SGOPS, and is essential to complement the goals of reliability studies. Functional interplay of various constituents of SG technologies must be captured for quantitative reliability assessment, to which end use-case based approaches are recommended. The chapter also includes some observations on reliability worth of investments in SGOPS, which are expected to figure in policy and regulatory legislations. Reliability-based appraisal of SG challenges eventually points to the recognition of an imminent need to deal with system reliability as an objective rather than a constraint, and is crucial to the successful realization of SG visions.

References

1. http://smartgrid.epri.com/. Accessed 16 Feb 2014
2. National Energy Technology Laboratory (2007) A systems view of the modern grid. https://www.smartgrid.gov/sites/default/files/pdfs/a_systems_view_of_the_modern_grid.pdf. Accessed 16 Feb 2014
3. Moslehi K, Kumar R (2010) A reliability perspective of the smart grid. IEEE Trans Smart Grid 1(1):57–64
4. Bose A (2010) Models and techniques for the reliability analysis of the smart grid. In: Proceedings of the IEEE PES General Meeting, Minneapolis, USA, pp 1–5
5. NERC Report (2010) Reliability considerations from the integration of smart grid. http://energy.gov/sites/prod/files/oeprod/DocumentsandMedia/SGTF_Report_Final.pdf. Accessed 16 Feb 2014
6. International Energy Agency (IEA) (2011) Technology roadmap: smart grids. http://www.iea.org/publications/freepublications/publication/smartgrids_roadmap.pdf. Accessed 16 Feb 2014
7. Vadlamudi VV, Karki R (2012) Reliability-based appraisal of smart grid challenges and realization. In: Proceedings of the IEEE PES General Meeting, San Diego, USA, pp 1–7
8. Farahmand H, Doorman G (2012) Balancing market integration in the northern European continent. Appl Energy 96:316–326
9. Zhang R, Zhao Z, Chen X (2010) An overall reliability and security assessment architecture for electric power communication network in smart grid. In: Proceedings of the international conference of the power system technology, Hangzhou, China, pp 1–6.
10. Wang Y, Li W, Lu J (2010) Reliability analysis of wide-area measurement system. IEEE Trans Power Deliv 25(3):1483–1491
11. Konig J, Franke U, Nordstrom L (2010) Probabilistic availability analysis of control and automation systems for active distribution networks. In: Proceedings of the IEEE PES transmission and distribution conference and exposition, New Orleans, USA, pp 1–8
12. Jensen M, Sel C, Franke U, Holm H, Nordstrom L (2010) Availability of a SCADA/OMS/DMS system—a case study. In: Proceedings of the IEEE PES European conference on innovative smart grid technologies, Gothenburg, Sweden, pp 1–8
13. Luan SW, Teng JH, Chan SY, Hwang LC (2010) Development of an automatic reliability calculation system for advanced metering infrastructure. In: Proceedings of the 8th IEEE international conference industrial informatics, Osaka, Japan, pp 342–347.
14. http://smartgrid.epri.com/Repository/Repository.aspx. Accessed 16 Feb 2014
15. http://ec.europa.eu/energy/gas_electricity/smartgrids/doc/2011_03_01_mandate_m490_en.pdf. Accessed 16 Feb 2014
16. Rinaldi SM, Peerenboom JP, Kelly TK (2001) Identifying, understanding, and analyzing critical infrastructure interdependencies, IEEE Control Systems Magazine, pp 11–25
17. Buldyrev SV, Parshani R, Paul G, Stanley HE, Havlin S (2010) Catastrophic cascade of failures in interdependent networks. Nature 464:1025–1028
18. Kjølle G, Utne IB, Gjerde O (2012) Risk analysis of critical infrastructures emphasizing electricity supply and interdependencies. Reliab Eng Syst Saf 105:80–89
19. Huiling S, Yang L (2011) Vulnerability control for power system by smart demand response. In: Proceedings of the Asia-Pacific power and energy engineering conference, Wuhan, China, pp 1–4
20. Lee KW, Tillman FA, Higgins JJ (1988) A literature survey of the human reliability component in a man-machine system. IEEE Trans Reliab 37(1):24–34
21. Billinton R, Allan RN (1984) Reliability evaluation of power systems. Plenum Press, New York
22. Adzanu SK (1998) Reliability assessment of non utility generation and demand side management in composite power systems, Ph.D. dissertation, Department of Electrical and Computer Engineering, University of Saskatchewan, Saskatoon, Canada

23. Kjølle G, Sand K (1992) RELRAD—An analytical approach for distribution system reliability assessment. IEEE Trans Power Deliv 7(2):809–814
24. U. S. Department of Energy Report (2007) The potential benefits of distributed generation and rate-related issues that may impede their expansion. http://energy.gov/sites/prod/files/oe-prod/DocumentsandMedia/1817_Report_-final.pdf. Accessed 16 Feb 2014
25. Costa PM, Matos MA (2010) Capacity credit of microgeneration and microgrids. Energy Policy 38(10):6330–6337
26. Chowdhury AA, Islam SM (2007) Development and application of probabilistic criteria in value-based transmission system adequacy assessment. In: Proceedings of the Australasian Universities power engineering conference, Perth, Australia, pp 1–9
27. Kjølle G, Samdal K, Singh B, Kvitastein OA (2008) Customer costs related to interruptions and voltage problems: methodology and results. IEEE Trans Power Syst 23(3):1030–1038
28. Vadlamudi VV (2011)Power system reliability-based techno-economic studies in the liberalized scenario, Ph.D. dissertation, Department of Electrical Engineering, Indian Institute of Technology Bombay, Mumbai, India.
29. http://www.smartgrids.no/. Accessed 16 Feb 2014
30. http://www.gridwiseac.org/pdfs/reliability_interoperability.pdf. Accessed 16 Feb 2014
31. https://standards.ieee.org/findstds/standard/2030-2011.html. Accessed 16 Feb 2014
32. http://www.nist.gov/smartgrid/upload/NIST_Framework_Release_2-0_corr.pdf. Accessed 16 Feb 2014
33. Gjerde O, Kjølle GH, Detlefsen NK, Brønmo G (2011) Risk and vulnerability analysis of power systems including extraordinary events. In: Proceedings of IEEE PES Powertech, Trondheim, Norway, pp 1–6
34. Doorman G, Uhlen K, Kjølle GH, Huse ES (2006) Vulnerability analysis of the Nordic power system. IEEE Trans Power Syst 21(1):402–410
35. Kjølle GH, Gjerde O, Hoffman M (2012) Monitoring vulnerability in power systems: extraordinary events, analysis framework and development of indicators. In: Proceedings of the 12th international conference on probabilistic methods applied to power systems, Istanbul, Turkey, pp 935–940

Security of Supply in Active Distribution Networks with PHEV-Based Strategic Microgrids

Dilan Jayaweera and Syed Islam

1 Introduction

Renewable power is a strong resource of green energy for the future of electricity generation. Among renewable energy sources, wind energy is the world's fastest growing energy source with an average annual growth rate of 29 % over the last 10 years. It is estimated that worldwide installed capacity of wind will reach 460 GW by 2015. Unlike conventional power plant sites, wind plants are geographically constrained and operated as geographically constrained intermittent power generating units that randomly suffer from network constraint issues in transporting electricity to destinations. Such problems can be mitigated by strategically placing the units that compensate for intermittent power outputs. One of such units is the plug-in hybrid electric vehicle (PHEV) reserve-based microgrid. Strategically placing PHEV reserve-based microgrids can potentially mitigate severity of network problems and they can also improve the security of supply if they are appropriately located in an active distribution network. Such a provision can also improve the active network performance by reducing power losses, by improving voltage profiles, by reducing the network congestion, and by improving power transfer margins. However, the challenging issue in this context is how to determine the locations of PHEV reserve-based strategic microgrids and how to quantify their capacities with the presence of increased renewable power generation. The challenge and the complexity of the problem are further increased when the overall objective is to improve the security of supply or at least to maintain the acceptable level of security of supply to consumers.

The integration capacity of distributed generation (DG) on existing distribution networks are limited and high penetration of DG can also harm the security of power supply to consumers unless the high penetration is made through strategi-

D. Jayaweera (✉) · S. Islam
Department of Electrical and Computer Engineering, Curtin University, Perth, Australia
e-mail: d.jayaweera@curtin.edu.au

S. Islam
e-mail: s.islam@curtin.edu.au

R. Karki et al. (eds.), *Reliability Modeling and Analysis of Smart Power Systems,*
Reliable and Sustainable Electric Power and Energy Systems Management,
DOI 10.1007/978-81-322-1798-5_2, © Springer India 2014

cally gifted locations. The intermittent power output characteristics of centrally integrated DG units such as wind and photo voltaic (PV) units have caused problems to the security of power supply at some of the penetration levels and the literature addresses these issues in detail [1–3]. Reference [4] proposes a linearized technique to quantify the risk-based index for the assessment of dynamic security. DG impacts can be viewed through many avenues. Reference [5] proposes a successive elimination algorithm to assess impacts of DG on investment deferral through the security of supply provision. Reference [6] presents an investment deferral approach based on demand growth and investigates the associated influence of the investment on security of supply. Impacts of large scale integration of wind power on network security through contingency analysis are explored in [7]. Optimal power flow and active management techniques-based approach is proposed in [8] for the economical assessment of wind-diesel connected systems. Reference [9] proposes a technique to assess impacts of location of wind farms. A method to determine the accommodation capacity of DG with the objective of minimizing the power losses with smart control switches is explored in [10]. The reference [11] investigates the PHEV applications in the deregulated electricity market and applies dynamic programming to optimize the charge control. Reference [12] assesses electric energy and power consumption with light duty plug in hybrid vehicles. Impacts of PHEV charging patterns are studied in [13] in the context of stochastic unit commitment.

This chapter investigates the role of PHEV-based microgrids for improving the security of supply to customers with the increased penetration of intermittent DG as the form of remote microgrids. The chapter also proposes a methodology to determine locations of the PHEV-based strategic microgrids and their capacities taking into account the mobility advantage of PHEV-based microgrids. The chapter focuses on what power the PHEV and remote microgrids deliver to an active distribution network and not the inside operation of microgrids. The steady state security is the focus of the approach which incorporates Monte Carlo simulation (MCS) to estimate the parameters.

The chapter is organized as follows. Section 2 presents the approach of the chapter. Section 3 describes the case study in detail and critically scrutinizes the results. Section 4 concludes findings.

2 The Approach

Outages in a distribution network can stress the nominal operating characteristics of equipment. Some of the active networks are congested with increased penetration of DG. In those circumstances, although the electricity can be generated at remote stations they may not always be able to transfer to the load centers. The severity of the situation can also be increased with the increased penetration of intermittent DG. Thus, identifying PHEV-based strategic microgrids and then penetrating renewable power through remote microgrids can be a potential solution to mitigate network congestion at stressed operating conditions. In addition, the PHEV-based strategic microgrids help reducing power losses as well as reduce the need for reactive power due to intermediate supply of active power. The most important aspect of a PHEV-based microgrid is the ability to vary its capacity dynamically taking into account the advantage of PHEV capacity mobilization.

The proposed approach incorporates sequential MCS to capture uncertainties in active distribution network operating conditions. Sequential MCS is used because the successor sample operating condition of MCS depends on the predecessor sample operating condition. In an active distribution network, most uncertainties are due to random outages of network components, intermittent power outputs from stochastic DG plants, and significant demand level variations. The severity of outages can also be increased with different weather conditions.

The proposed approach treats the wind and PV plants as generating sources of remote microgrids. Thus, the strategic PHEV-based microgrids are able to provide standing reserve supports for the intermittent outputs of remote microgrids and help reducing the network stress due to outages and congestion.

The first step of the approach is to model the base case operating condition without incorporating PHEV-based microgrids. Adopting this philosophy enables to differentiate the impacts of intermittent DG-based microgrids from PHEV-based microgrids. The base case should be selected in such a way that it does not carry constraint violations. Constraints in this assessment include thermal limits of branches and voltage limits of buses.

Next, remote microgrids with wind and PV generating units are identified and they are connected to the active distribution network. These microgrids can inject power to the active distribution network. Then, the network component outages are modeled by generating random numbers and comparing with probability of failure of network components. For example, if a component has a probability of failure of α and the generated random number corresponding to the network component is β then satisfying the condition $\alpha > \beta$ sets the component status as in service. If this inequality condition is not satisfied then the state is considered as out of service.

In this way, the network topological arrangement is determined for the sample and the system equilibrium is assessed by running A/C power flow algorithm. Convergence of power flow can also result constraint violations. Any violation is rectified by applying active network management (ANM) controls. ANM controls include dispatching of flexible generation, on load tap changing (OLTC) of transformers, shunt compensation, and load shedding. The load shedding is applied only if other ANM controls failed to rectify the constraint violations or the resulting state following the application of power flow algorithms results an unstable operating condition. The order of the application of ANM control is followed as re-dispatching of flexible generation, OLTC of transformers, shunt compensation, and at last load shedding if that option is compulsory and appropriate. These control actions are sequentially applied and the sensitivity of the ANM control to reduce the constraint violations is determined. In this way, one or more ANM is applied to rectify any constraint violation. Upon rectifying all violations, the energy not supplied (ENS) of the sample of MCS is calculated using the amount of shed load and time to restore it. ENS is defined as the product of the amount of shed load and restoration time of the shed load.

Next, the PHEV-based microgrids are connected to the distribution network through feasible PHEV sites. PHEV microgrid reserve of each site is quantitatively injected to the network and the sensitivity of each PHEV-based microgrid reserve to reduce the ENS value (resulted through the disturbances of the sample) is assessed. At this stage two options are proposed. The first option considers simultaneous but proportional injection of active power from PHEV-based microgrids to reduce the ENS and calculates the active power capacity of necessary PHEV from each PHEV microgrid site when they are operated simultaneously.

The second option considers stand-alone injection of active power to the distribution grid or in other words, PHEV-based microgrids inject active power into the distribution network on one at a time basis and calculates the active power capacity of necessary PHEV if the PHEV microgrids are to operate as stand-alone microgrids. Both options can have benefits depending on the probability of failure of components surrounding the PHEV-based microgrids in the active distribution network. For example, given that no outage affects the availability of any PHEV microgrid then option 1 can give most beneficial PHEV dispatch. The option 2 is beneficial if one concerns of out of service of any PHEV microgrid that affects the duty of PHEV-based microgrids to reduce ENS. In this way, the ENS of a sample of MCS is calculated and the level of reduction in ENS obtained through PHEV-based microgrid deployment is determined.

The process described above occurs in a sample of MCS. The process continues until the MCS meets the convergence criteria [14]. Three conditions are set to meet the convergence criteria. It should first run trials that can at least capture the time series profiles of intermittent DG and demand level variations in a year. Then, the parameter to be estimated (curtailed load) is used to check if it meets the degree of confidence within an acceptable confidence interval, which is also called as tolerance criterion. It can mathematically be interpreted as in Eq. (1).

$$\sigma \leq \frac{E \times L}{\alpha} \tag{1}$$

where, E, L, σ, and α are estimated sample mean, confidence interval, estimated sample standard deviation, and the factor that represent postulated normal distribution for the accepted degree of confidence respectively. If both of above criteria were not met, then MCS is stopped by meeting the maximum number of trials. With the convergence of the MCS the expected energy not served (EENS) is calculated using Eq. (2).

$$EENS = \frac{1}{n}\sum_{i=1}^{n} P_i \times T_i \tag{2}$$

where n = number of samples processed, P_i = the magnitude of curtailed load at the sample i, and T_i = time to restore P_i. Figure 1 shows the basic steps of the approach in a sample of the MCS that is used to determine the strategic microgrids.

The ENS in a sample can also be interpreted with the cost of outage using value of lost load functions [15]. With the convergence of the MCS, the expected cost of outage (ECO) can be estimated using Eq. (3).

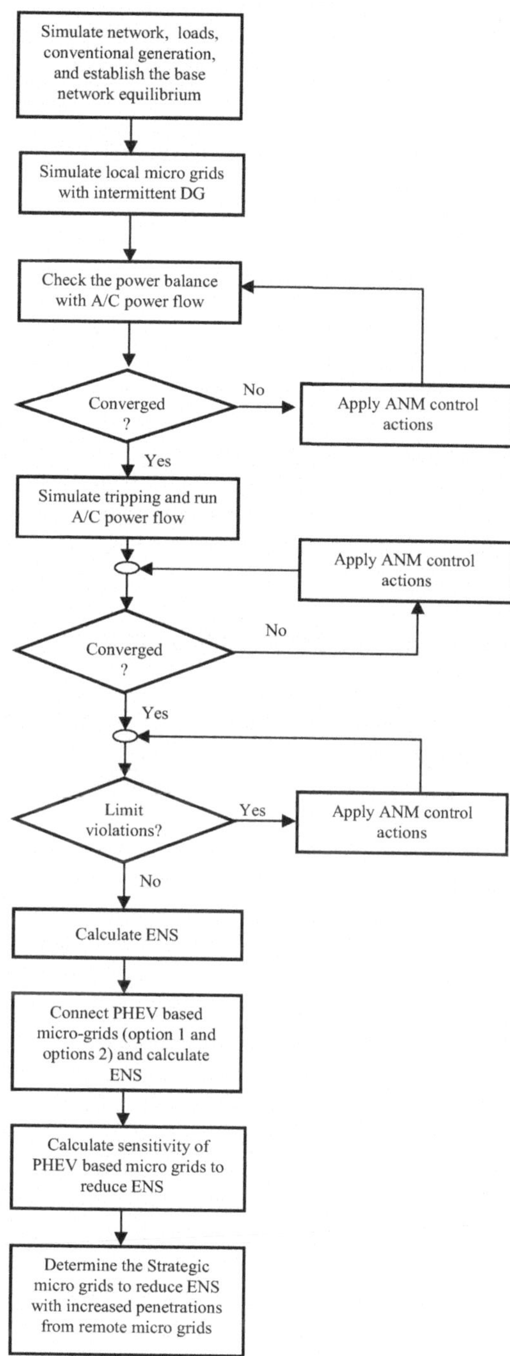

Fig. 1 Determination of PHEV-based strategic microgrids in a sample in MCS

$$ECO = \frac{1}{n}\sum_{i=1}^{n}\left(VoLL_i^t \times ENS_i\right) \quad (3)$$

where $VoLL_i^t$ and ENS_i respectively referred to value of lost load for the interruption duration t and ENS for the sample i.

2.1 Ranking Remote Microgrids

The remote microgrids are used to increase the intermittent DG penetration. They are connected to the active distribution network on the basis of one at a time. The process described above is repeated and the EENS is calculated for each of the penetration level. In this way, all the remote microgrids can be ranked in ascending order of reducing EENS with high penetration of intermittent DG. The most favorable remote microgrid can be identified as the microgrid that reduces EENS most while increasing the penetration level of intermittent DG.

2.2 Strategic PHEV Microgrids

The active power capacities of PHEV-based microgrids are determined as follows. At each sample of the MCS the net MW import/export from each PHEV-based microgrid is calculated. In some samples, it can be seen as the exporting active power as the favorable option to reduce ENS whereas in other samples it can be seen as the importing MW as the favorable option to reduce ENS. Here, exporting and importing respectively referred to discharging and charging of all PHEVs in a microgrid. In this way, the exporting/importing of net active power through PHEV-based microgrids are calculated and recorded. Upon convergence of the MCS, the required net active power capacities of PHEV-based microgrids are determined by identifying the capacity that would fulfill all the net MW import and export of samples of MCS. The term “required active power” is used in this chapter to describe the useful or effective capacity of a PHEV-based microgrid because there is a threshold capacity level of PHEVs that is useful in exporting/importing powers. As the one of the objectives of the chapter is to assess the level of security with the presence of remote and PHEV-based microgrids, the chapter considers what PHEV-based microgrids deliver and what PHEV-based microgrids absorb from an active distribution network or in other words net in and out of microgrids.

The PHEV-based microgrids that reduce or at least maintain the base case level of EENS (i.e., resulting EENS value without the connection of intermittent DG and PHEV-based microgrids) by responding to increased penetrations of intermittent DG can be identified as strategic microgrids of the active distribution network.

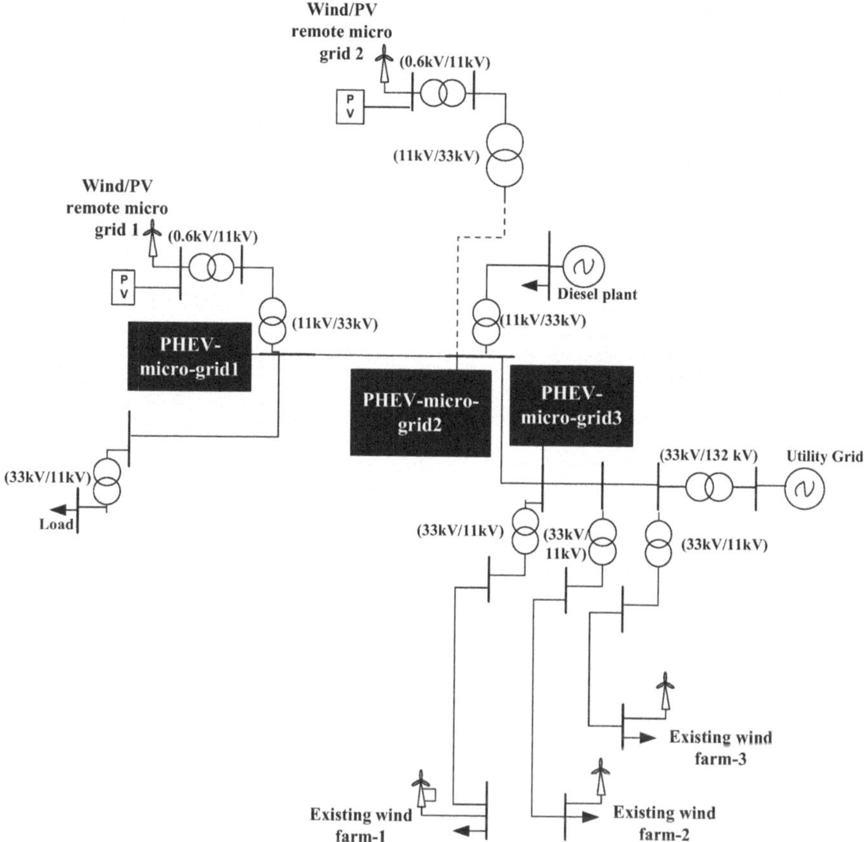

Fig. 2 Schematic diagram of a model of an active distribution network

3 Case Studies

3.1 Network Details

Figure 2 shows an active distribution network model that is used for the case studies. The network has 17 buses, 8 transformers, and 8 lines.

The 132/33 kV transformer connected at the utility bus has the OLTC operation. Remote microgrids shown in Fig. 2 represent only the power export operation under the grid connected mode. The microgrids supplying their own local demands are not simulated. Similarly the PHEV-based microgrids supplying their own local demands are not simulated. PHEV charging and discharging operations are simulated as net import and export of active power. For example if PHEV charging cycle or discharging cycle reduces the ENS of a sample of MCS then corresponding cycle is taken into account as a component of sizing the net capacity of PHEV-based microgrids.

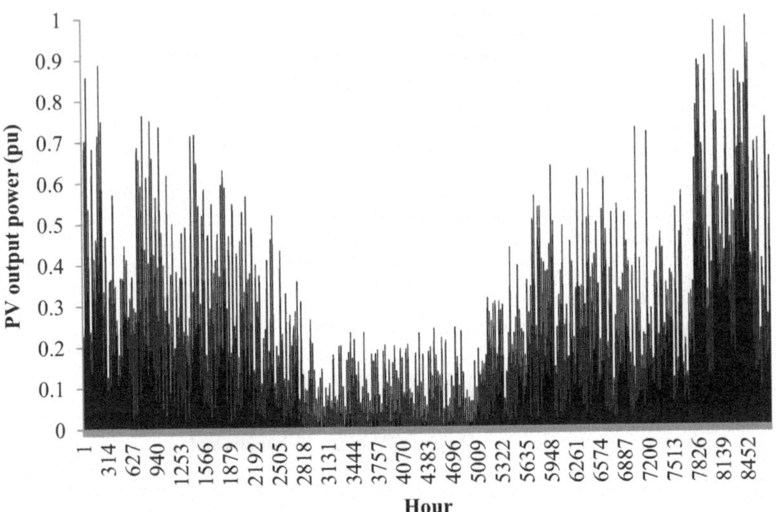

Fig. 3 Photo-voltaic (PV) output characteristics in a year

The total peak active and reactive power demands of the base case of the network are 15.2 MW and 5 MVAr respectively. There are two remote microgrid sites that can be connected to the active distribution network. There are also three proposed PHEV-based microgrids that are shown in Fig. 2 as PHEV microgrid 1, PHEV microgrid 2, and PHEV microgrid 3. Installed capacity of wind and PV units of each of the remote microgrid are 2 MW and 0.6 MW respectively. Each of the existing wind farm installed capacity is 0.7 MW. Intermittency of wind and PV units are modeled using hourly wind and PV power generation profiles. Figure 3 and 4 respectively show the PV and Wind data for a year respectively. There are three types of customers simulated for the assessment. They are residential, industrial, and commercial. The wind farm capacity factor accounts for 0.27 whereas PV system capacity factor accounts for 0.12.

3.2 *Scenarios*

A set of scenarios are developed to study the performance of the proposed approach and to determine the PHEV-based microgrid capacity that can increase the integration capacity of intermittent DG without compromising the security of supply to customers. Each scenario used for the study is described below.

Scenario 1 Base case which does not have remote or PHEV-based microgrids. Thus, intermittent wind and PV generating units and PHEV reserves are not present in the active distribution network in Scenario 1.

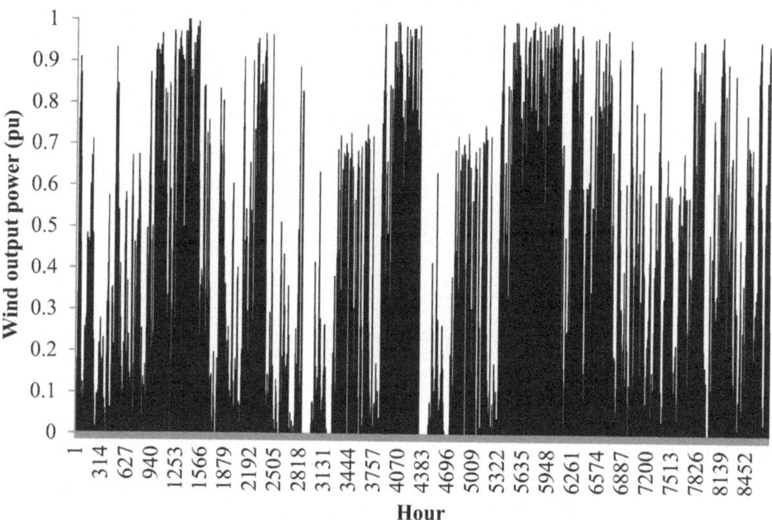

Fig. 4 Wind generation output characteristics in a year

Scenario 2 Base case with remote microgrids but without the integration of PHEV-based microgrids. Thus, the active distribution network is integrated with intermittent wind and PV power generation without PHEV-based microgrid supports.

Scenario 3 Base case with PHEV-based microgrids are connected at three locations in the active distribution network. These locations are shown in Fig. 2 as PHEV microgrid 1, PHEV microgrid 2, and PHEV microgrid 3. Network is also connected with remote microgrids as described in scenario 3(a) and 3(b).

Scenario 3(a) Remote microgrid 1 which has wind and PV generating units.

Scenario 3(b) Remote microgrid 2 which has wind and PV generating units.

3.3 Results and Analysis

Figure 5 shows EENS values of the base case with and without incorporating remote microgrid1 that exports intermittent renewable power to the active distribution network. The results depict that the component outages and existing 2 MW wind farm combinatorial impacts result an EENS of 10.3 MWh annually whereas injecting a further renewable power of 2.7 MW reduces the EENS value to 9.6 MWh annually. Thus, the reduction of 0.7 MWh is directly a result of the influence of 2.7 MW renewable power injection from the remote microgrid 1.

Figure 6 shows the required active power capacity of PHEV-based microgrid1, 2, and 3 to have the EENS values shown in Fig. 7. In other words, the required ac-

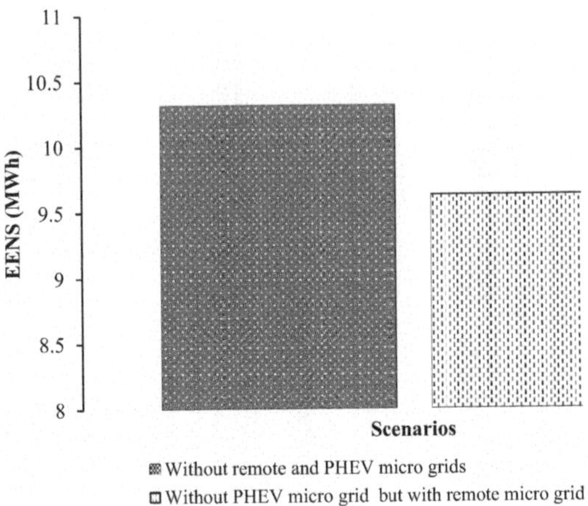

Fig. 5 EENS values with base case operating conditions and incorporating remote microgrid 1

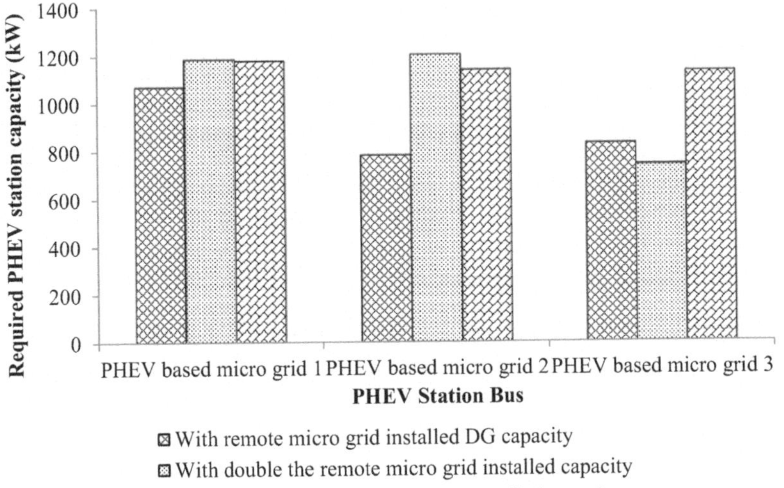

Fig. 6 Required PHEV supports from respective PHEV-based microgrids in response to increased penetration levels of wind and PV generation from remote microgrid 1

tive power capacity represents the necessary installed capacity of the PHEV-based microgrids to meet the net active power import/export of all MCS samples. When the remote microgrid 1 injects power to the distribution network with three times the remote microgrid installed capacity, all PHEV-based microgrids have similar sensitivity in reducing EENS values although the PHEV microgrid 3 can mostly reduce the EENS. However, when the required active power capacity of PHEV-based microgrid and penetrating renewable generation is compared with the initially in-

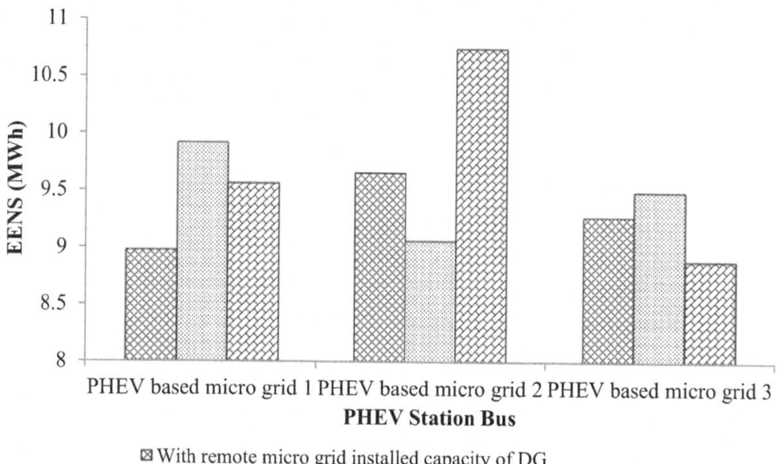

Fig. 7 EENS values with the incorporation of PHEV microgrids and increased penetration levels of wind and PV generation through the remote microgrid 1

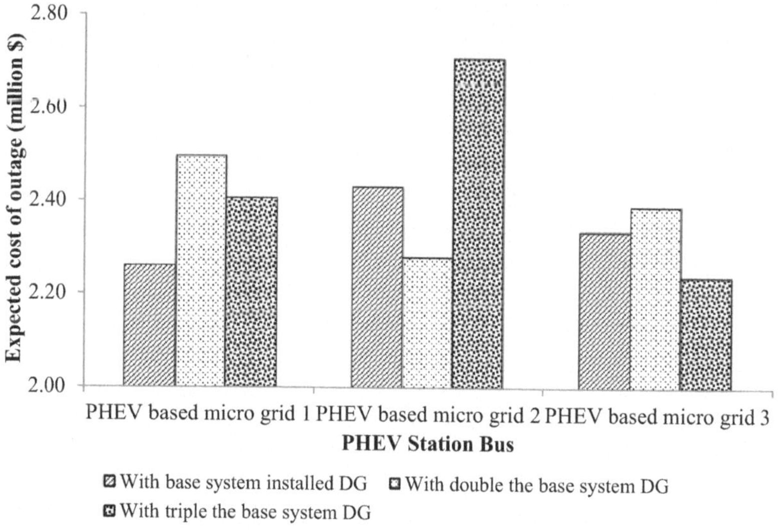

Fig. 8 ECO for EENS values in Fig. 7

stalled capacity of remote microgrids, PHEV microgrid 2 is the most favourable option due to its strategic ability to improve the security of supply. When the renewable power penetration is doubled compared to the initially installed capacity then the most favourable PHEV-based microgrid, for this particular network, would be

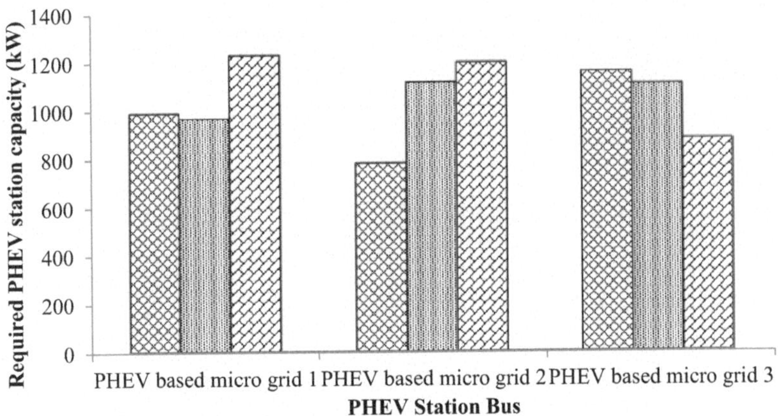

Fig. 9 Required PHEV supports from PHEV-based microgrids for the increased penetration levels of wind and PV power from remote microgrid 2

the PHEV microgrid 3. Figure 8 shows the ECO corresponding to EENS values in Fig. 7. They justify the significance of security of supply to electricity consumers in million dollars.

Figure 9 shows the required PHEV active power capacity to have EENS values shown in Fig. 10. In this scenario, the remote microgrid 2 is connected to the active distribution network. When compared the required active capacity in Fig. 6 with Fig. 9, it is clear that the penetrating renewable power from the remote microgrid 2 offers a reduced demand of active power capacity from PHEV-based microgrids. In terms of reducing EENS, the PHEV-based microgrid 1 provides the highest reduction in EENS of all the penetartion levels of remote microgrid 2.

Figure 11 shows the ECO of EENS values in Fig. 10. Results suggest that a significant variation in ECO exists when the Wind and PV are penetrated through the remote microgrid 2.

As whole, the case study results suggest that PHEV-based microgrids can provide strategic supports and reduce EENS or in other words they are able to improve or maintain the security of supply to customers with the increased penetration of intermittent DG through remote microgrids. There is no single PHEV-based microgrid that can constantly provide strategic supports for all the penetration levels of renewable power penetrations through remote microgrids. The strategic status of PHEV-based microgrids dynamically varies with the varying level of penetrations from remote microgrids. Thus, it is apparent that the proposed approach will be a reality in the context of smartgrids.

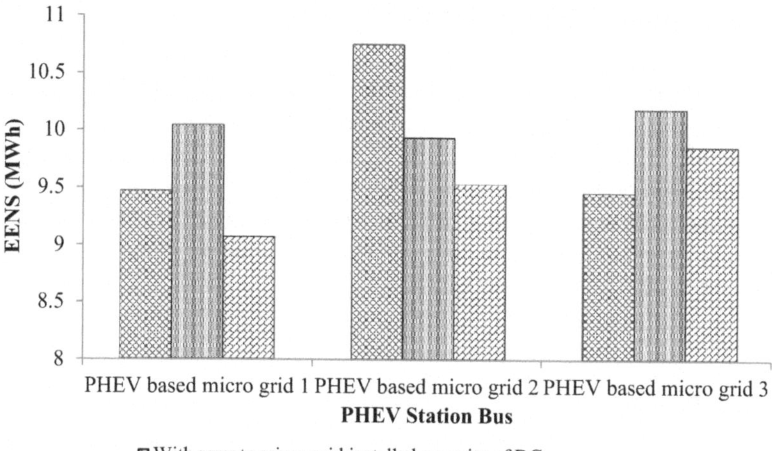

Fig. 10 EENS value with the incorporation of PHEV and increased penetration levels of wind and PV generation through the remote microgrid 2

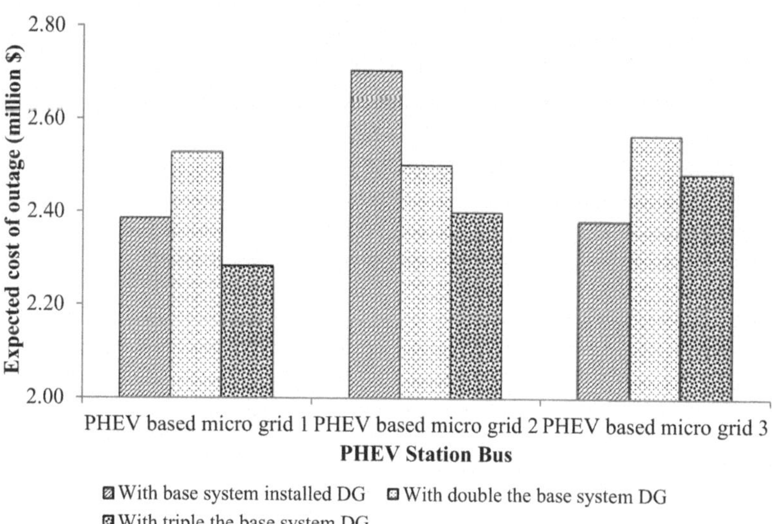

Fig. 11 ECO for EENS values in Fig. 10

4 Application into Smartgrid

In the smartgrid context, smart sensors can be embedded in PHEV-based and remote microgrids. An intelligent electronic control module which may be located in the active distribution network can be trained with the proposed algorithm. The module can then communicate with PHEV-based and remote microgrids and net charging and discharging cycles can be monitored dynamically. Thus, the module can provide periodic information on the level of intermittent DG that can be embedded into the active distribution network.

5 Conclusions

An approach is proposed to assess the level of PHEV microgrid supports to improve or at least to maintain the security of supply to customers with the increased penetrations of intermittent DG.

The performed case study results suggest that there is a considerable potential to increase the intermittent DG with the strategic placement of PHEV-based microgrids. Studies also suggest that PHEV-based microgrids can provide strategic supports for the accommodation of increased amount of intermittent DG. However, the strategic support microgrid is not necessarily the same PHEV-based microgrid for all the penetration levels of intermittent DG. The strategic status of PHEV-based microgrids can dynamically vary if the intermittent DG penetration is increased.

The results also argue that there is a need for dynamic coordination of PHEV-based microgrids with the increased penetration of DG. With the advances in smartgrid technologies, the dynamic coordination facilities can be a reality in active distribution networks. The proposed approach can also be used as one of benchmarking entity for the assessment of improved security in smart grids.

Acknowledgments The authors gratefully acknowledge the financial support provided to this research by the CSIRO Flagship Cluster on Intelligent Grid—Distributed Generation.

References

1. Jayaweera D, Burt G, McDonald D (2007) Customer security assessment in distribution networks with high penetration of wind power. IEEE Trans Power Syst 22:1360–1368
2. Hegazy YG, Salama MMA, Chikhani AY (2003) Adequacy assessment of distributed generation systems using Monte Carlo simulation. IEEE Trans Power Syst 18:48–52
3. Jayaweera D, Islam S (2009) Probabilistic assessment of distribution network capacity for wind power generation integration, in Power Engineering Conference (AUPEC 2009), Adelaide, South Australia, pp. 1–6
4. Dissanayaka A, Annakkage UD, Jayasekara B, Bagen B (2011) Risk-based dynamic security assessment . IEEE Trans Power Syst 26:1302–1308

5. Wang DTC, Ochoa LF, Harrison GP (2010) DG impact on investment deferral: network planning and security of supply. IEEE Trans Power Syst 25:1134–1141
6. Wang DT, Ochoa LF, Harrison G (2010) DG impact on investment deferral: network planning and security of supply. IEEE Trans Power Syst 25:1134–1140
7. Grijalva S, Dahman SR, Patten KJ, Visnesky AM (2007) Large scale integration of wind generation including network temporal security analysis. IEEE Trans Energy Conv 22:181–188
8. Siano P, Citro C, Cecati C, Piccolo A (2011) Smart operation of wind turbines and diesel generators according to economic criteria. IEEE Trans Ind Electron 99:1
9. Hamidi V, Li F, Yao L (2011) Value of wind power at different locations in the grid. IEEE Trans Power Deliv 26:526–537
10. Ochoa LF, Harrison GP (2011) Minimizing energy losses: optimal accommodation and smart operation of renewable distributed generation. IEEE Trans Power Syst 26:198–205
11. Rotering N, Ilic M (2011) Optimal charge control of plug-in hybrid electric vehicles in deregulated electricity markets. IEEE Trans Power Syst 26:1021–1029
12. Wu D, Aliprantis DC, Gkritza K (2011) Electric energy and power consumption by light-duty plug-in electric vehicles. IEEE Trans Power Syst 26:738–746
13. Liu C, Wang J, Botterud A, Zhou Y, Vyas A (2012) Assessment of impacts of PHEV charging patterns on wind-thermal scheduling by stochastic unit commitment. IEEE Trans Smart Grid 3:675–683
14. Kirschen DS, Jayaweera D, Nedic DP, Allan RN (2004) A probabilistic indicator of system stress. IEEE Trans Power Syst 19:1650–1657
15. Kariuki KK, Allan RN (1996) Evaluation of reliability worth and value of lost load. IEE Proc Gener Trans Distrib 143:171–180

Operational Characteristics of Microgrids with Electric Vehicles

Clara Sofia Gouveia, Paulo Ribeiro, Carlos L. Moreira and João Peças Lopes

1 Introduction

The development of a competitive electric power sector associated with environmental and security of supply issues have led to a new vision of future electricity networks—the smart grid concept. Under this new paradigm, the Microsources (MS) and the electrical vehicles (EV) connected to low voltage (LV) distribution networks can be actively managed exploiting the microgrid (MG) structure [6, 9]. The MG is a highly flexible, active, and controllable LV cell, incorporating local generation resources based on renewable energy sources (RES) or low carbon technologies for combined heat and power (CHP) applications, storage devices, and loads, which are controlled and coordinated though an appropriate network of controllers [5, 10]. The MG management and control system extends and decentralizes the distribution network monitoring and control capability, providing the adequate framework to fully integrate smart grid new players, namely microgeneration, EV, and responsive loads [5, 6, 10, 11].

Being an extremely flexible cell of the electrical power system, the MG is able to operate interconnected to the main power system (normal operation) or autonomously, when major disturbances occur in the upstream network (emergency operation) [10]. This new distribution network operation philosophy increases the system reliability and resilience against component failures and natural disasters

C. S. Gouveia (✉) · P. Ribeiro · C. L. Moreira · J. P. Lopes
INESC TEC—INESC Technology and Science (formerly INESC Porto) Porto, Campus da FEUP, Rua Dr. Roberto Frias, 378, 4200–465, Porto, Portugal
e-mail: cstg@inescporto.pt; clara.s.gouveia@inescporto.pt

P. Ribeiro
e-mail: pjsr@inescporto.pt

C. L. Moreira
e-mail: cmoreira@inescporto.pt

J. P. Lopes
e-mail: jpl@fe.up.pt

R. Karki et al. (eds.), *Reliability Modeling and Analysis of Smart Power Systems*, Reliable and Sustainable Electric Power and Energy Systems Management, DOI 10.1007/978-81-322-1798-5_3, © Springer India 2014

[5]. However, ensuring a successful MG autonomous operation is quite challenging. In the moments subsequent to the islanding transient, the MG can suffer severe frequency and voltage disturbances. In order to assure system survival, the MG stability relies in local voltage and frequency control strategies, exploiting energy storage devices, controllable MS, and flexible loads [10].

The deployment of EV may offer additional resources to the MG operating under emergency conditions [11, 12]. When properly managed, the EV can be regarded as a very flexible load or storage device: when parked and connected to the LV network through proper interfaces, EV will controllably absorb energy and store it in their batteries, being also able to deliver it back to the grid—the vehicle-to-grid (V2G) concept. The adoption of innovative control strategies, taking advantage of the EV flexibility, has the potential to enhance MG resilience, avoiding large frequency excursions resulting from loads or MS power variations [11, 12].

The MG autonomous operation cannot be evaluated without considering the unbalanced nature of the LV distribution systems. Voltage unbalance occurs mainly due to the uneven connection of single-phase loads and MS to the three phases of the system, decreasing lifetime of three-phase loads (such as motor loads) and compromising the MG synchronization with the upstream network. When operating in islanded mode, the MG is more sensitive to this problem, which could be further accentuated by the connection of single-phase EV charging interfaces [11]. Therefore, active voltage compensation strategies are required in order to eliminate the unwanted negative and zero sequence voltages components, mainly during MG islanded operation.

2 Microgrid Architecture

As shown in Fig. 1, the MG management and control system is organized in a network of local controllers with local intelligence headed by the MG central controller (MGCC), which coordinates all the MG resources and communicates with control systems responsible for managing the distributing grid at the medium voltage (MV) level.

The MGCC is installed at the MV/LV substation and concentrates the high-level decision making for the technical management of the MG. Its processing capabilities allow filtering and processing the information sent by the local controllers, which will be used not only by the MGCC but also sent to control systems at the MV level.

Considering the different MG actors, there are three types of local controllers, namely: the load controller (LC), the microsource controller (MC), and the EV controller (VC) [10, 11]. As proposed in [11, 12], the EV-dedicated controller enables the control of the EV battery charging rate and the exchange of information between the EV and the MGCC. Similarly, the controllers located at loads (LC) and the controllers located at MS (MC) exchange information with the MGCC that manages MG operation and the MGCC provides set-points to both LC and MC. LC constitutes the interface to control flexible loads and can also include local load

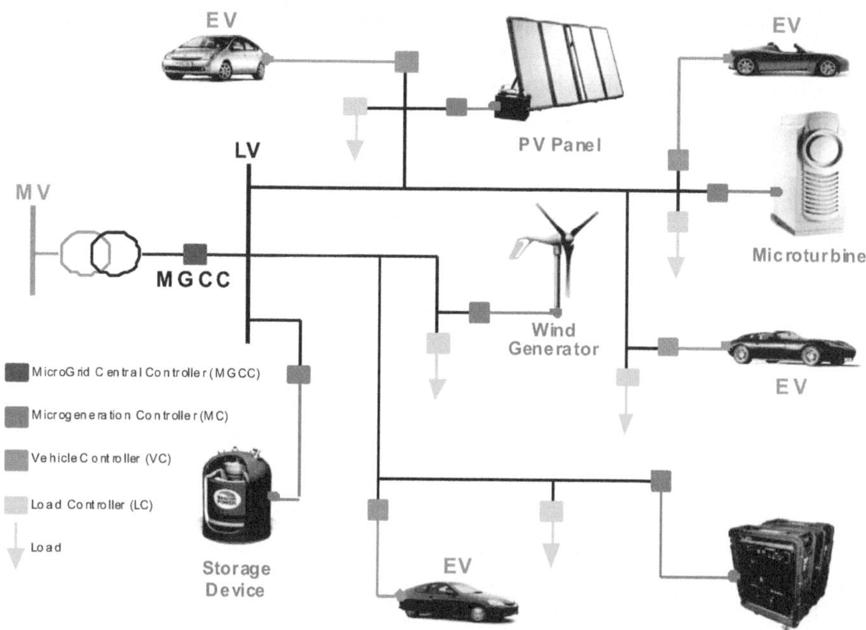

Fig. 1 MG architecture including EV [10, 11]

shedding schemes for improving MG operation during an emergency situations. MC control active and reactive power production levels at each MS, following the management profile defined by the MGCC.

In normal interconnected mode, the MGCC performs MS dispatch, voltage coordination, security assessment, and storage charging management, transmitting the resulting control set-points to the local controllers [4–6]. During islanded operation, the main objective is to maintain the MG stability and ensure a smooth operation. In these operating conditions, the MG relies mainly in its control architecture to reestablish the MG voltage and frequency to nominal values, avoiding the dependence from fast communication infrastructures. However, supervisory control is always necessary in order to adequately manage the available resources in a larger time frame [4, 10].

3 MG Regulation Functionalities for Emergency Operation

When operating autonomously, the MG requires the adoption of local frequency and voltage regulation strategies in order to assure system stability. Specific primary and secondary frequency control strategies have to be adopted, since the system is inertialess due to the inexistence of rotating masses connected to the MG.

The frequency regulation strategies are implemented by controlling the MS grid coupling inverter, considering the MG resources controllability, namely: the main storage device response, EV charging controllability, controllable MS response, and load shedding schemes [4, 10–12].

The MG primary frequency control is provided by storage units capable of providing fast power balance (i.e., flywheels, ultracapacitors, etc.). In order to provide frequency regulation, the storage unit grid coupling inverter is controlled as a voltage source inverter (VSI) device with external droop control loops, defining the MG voltage and frequency references as a function of the grid operating conditions, as in Eq. (1).

$$\begin{aligned} \omega &= \omega_0 - k_P \times P \\ V &= V_0 - k_Q \times Q \end{aligned} \tag{1}$$

where P and Q are the inverter active and reactive power outputs, k_P and k_Q are the droop slopes (positive quantities), and ω_0 and V_0 are the idle values of the angular frequency and voltage (values of the inverter angular frequency and terminal voltage at no load conditions).

The primary frequency control can be complemented by the implementation of load shedding schemes, in order to avoid large frequency excursions, which can compromise MG transient stability in the moments subsequent to islanding. The load shedding mechanism can be implemented as an emergency functionality in the LC, based on underfrequency relays.

Primary frequency control does not ensure the restoration of frequency to its nominal value, requiring the adoption of secondary frequency regulation strategies. Both local and centralized secondary control strategies have been proposed in the literature, considering the MG controllable MS (i.e., single-shaft microturbines—SSMT and solid oxide fuel cells—SOFC), which are dispatchable resources, operating with specified active and reactive power outputs. Local secondary frequency control is implemented at the controllable MS through PI controllers, as proposed in [10]. Centralized strategies at the MGCC level are used to define new PQ set-points for the controllable MS taking into consideration the overall MG operating state, in order to restore frequency and voltage magnitude to nominal values. Secondary control may also include additional synchronization loops for a smooth reconnection to the main grid after islanding [4].

3.1 Integrating EV in MG Emergency Operation

The load controllability and distributed storage capacity provided by the EV connected to the LV network are additional resources that can be exploited for MG frequency regulation purposes. In [11, 12], the authors propose a P-ω droop control strategy implemented at the EV charger, where the EV will modify the power exchange with the LV grid based on the MG frequency. As shown in Fig. 2, for fre-

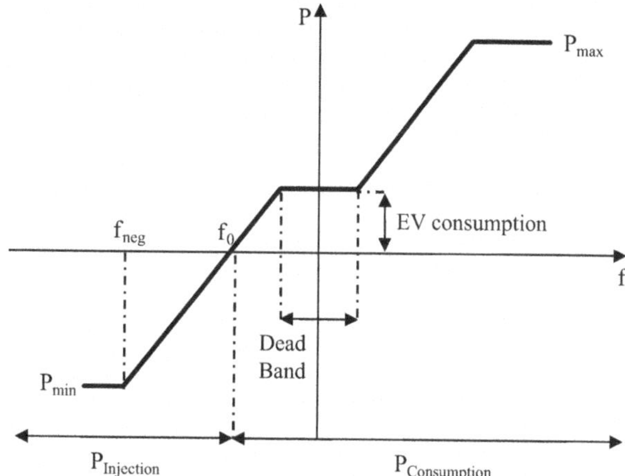

Fig. 2 EV frequency-droop characteristic

quencies around the nominal value (in this case, 50 Hz) the EV will charge the battery at a pre-defined charging rate. If a disturbance occurs and the frequency drops below the dead-band minimum, the EV reduces its power consumption, reducing the load of the system. When the MG frequency overpasses the frequency dead-band maximum, the EV can also increase its power consumption. For large disturbances, causing the frequency to go below the zero-crossing frequency (f_0), the EV starts to inject power into the grid (vehicle-to-grid (V2G) functionality). When the MG frequency becomes out of the predefined frequency range the vehicle will inject/absorb a fixed power.

The definition of the EV control parameters will depend on the EV charger characteristics and on the willingness of EV owners to participate in such services. These parameters may differ from grid to grid and can be changed remotely by the MGCC, in order to promote adequate coordination with the MG frequency regulation mechanisms (load shedding schemes, availability of energy storage devices, and their state of charge).

4 MG Unbalanced Operation

Due to the nature of a low voltage MG, the coexistence of single-phase connected loads and microgeneration, together with single-phase connected EV power electronic interfaces can significantly jeopardize power quality due to voltage imbalance problems. According to the European Standard EN50160, voltage unbalance can be measured by the voltage unbalance factor (VUF_{neg}) determined by Eq. (2), where V_1, V_2 are respectively the positive and negative sequence voltage values. In each bus the negative phase sequence component should be within 2 % of the positive phase sequence, given a 10-min average value [2]. The zero sequence voltage components have no significant impact on the MG loads. However, at the

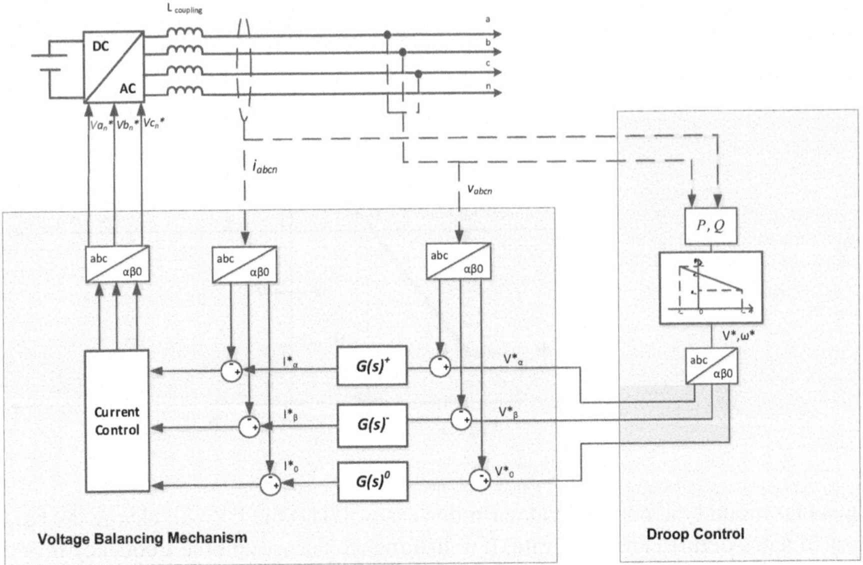

Fig. 3 Four-legged VSI with voltage balancing unit connected to the MG main storage unit

point of common coupling (PCC) it may compromise the synchronization of the islanded MG with the MV grid. Therefore, in this work the zero sequence voltage unbalance (VUF_{zero}) was also measured as indicated in Eq. (3), V_0 being the zero sequence voltage.

$$\%VUF_{neg} = \frac{V_2}{V_1} \cdot 100\% \tag{2}$$

$$\%VUF_{zero} = \frac{V_0}{V_1} \cdot 100\% \tag{3}$$

4.1 *Voltage Balancing Control Mechanisms*

In order to provide adequate voltage reference and eliminate the unwanted negative and zero sequence voltage, the voltage source inverter connected to the main storage can be complemented with an appropriated voltage balancing mechanism, being its main building block represented in Fig. 3. The major objective of the voltage balancing mechanism is to provide a three-phase balanced voltage at the VSI, regardless of the load current. The main control system of the VSI inverter with a voltage balancing control mechanism consists on a voltage-current regulation block and an external droop control block. In order to allow the independent control of the voltages in the three phases, the use of a four-legged inverter is required.

The VSI terminal voltage and current are measured in order to compute active and reactive powers. This measuring stage introduces a delay for decoupling purposes. The active power determines the frequency of the output voltage by the active power/frequency droop, and the reactive power determines the magnitude of the output voltage by the reactive power/voltage droop, as in Eq. (1).

The voltage-current regulation block is implemented in the stationary αβ0 reference frame. The use of the stationary reference frame is computationally more efficient, since it avoids frame transformations for the positive and negative voltage components [7, 13]. However, it requires the use of a specific controller, which is the resonant controller, in opposition to simple PI controllers that can be used when a synchronous reference frame control structure is adopted. The characteristic of this controller achieves a very high gain (tends to infinite) around the resonance frequency, which makes it capable of eliminating the steady state error between the controlled and the reference signal. However, the practical implementation of such controller is not feasible due to the resulting gains around the resonance frequency.

Therefore, a more practical implementation form can be written according to the following transfer function, which approximates the ideal integrator by Eq. (4) [7].

$$G(s) = K_p + \frac{K_i \omega_{cut} s}{s^2 + 2\omega_{cut} s + \omega^2} \tag{4}$$

where ω_{cut} is the low-frequency cutoff (being $\omega_{cut} \ll \omega$). The inner current control block (Fig. 3) is a proportional controller, which allows the derivation of the inwverter individual phase voltages from the inverter current references.

5 MG Dynamic Model

In order to characterize MG resilience in the moments subsequent to islanding considering the participation of EV, a full dynamic simulation model was developed. The LV network was modeled as a three-phase four-wire system, with both three-phase and single-phase loads and MS connected. Under unbalanced operating conditions, accurate power measurements are essential for robust operation of MG. The instantaneous active power of a three-phase circuit, p and the reactive (nonactive) power, q is calculated as in Eqs. (5) and (6).

$$p = v^{-T} . \bar{i} \tag{5}$$

$$q = \left\| q_{space} \right\| = \left\| \bar{v} \times \bar{i} \right\|. \tag{6}$$

For a single-phase connection, the power was measured according to the proposal of Burger and Engler [3]. With this approach, no zero crossing detection is required.

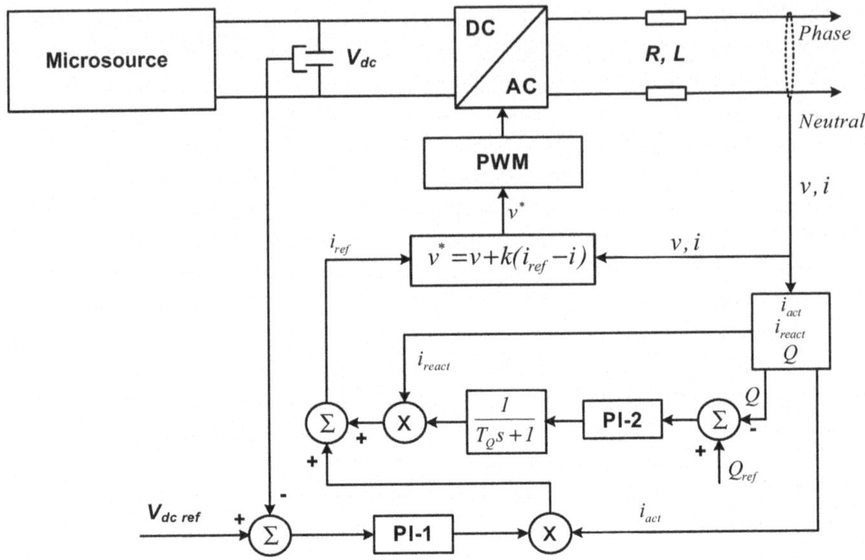

Fig. 4 Single-phase grid coupling inverter PQ control

The power is computed by using the complex apparent power split-up in its real and imaginary part, as in Eq. (7).

$$\underline{s} = p + jq = \frac{1}{2}\underline{v}.\underline{i}^* = \frac{1}{2}(v_r + jv_i)(i_r - ji_i) \tag{7}$$

where v_r and i_r are fictitious orthogonal components of the voltage and current, similar to the space vectors in three-phase systems.

The MS dynamic models have the following components, as represented in Fig. 4 [8, 10]:

- MS model (fuel cell, microturbine, PV panel, and micro-wind turbine).
- DC-link connected to the MS and to the DC-AC grid-coupling inverter.
- Coupling inductance.

A detailed description of the three-phase models of the controllable MS, such as SSMT and SOFC can be found in [8]. Due to the short-term nature of the studies to be performed, single-phase MS such as PV panels and micro-wind turbines were modeled as constant power sources.

The MS DC-AC grid-coupling inverters were modeled as current controlled voltage source (PQ controlled inverter, in general), as in Fig. 4. MS power variations will cause voltage error in the DC link which will be corrected through PI-1 and change the active current delivered to the grid. The reactive power can be controlled through PI-2.

Figure 4 represents the block diagram of a single-phase MS DC-AC grid-coupling inverter. Since the majority of single-phase MS are based on variable RES, the inverter will inject the power available at its input.

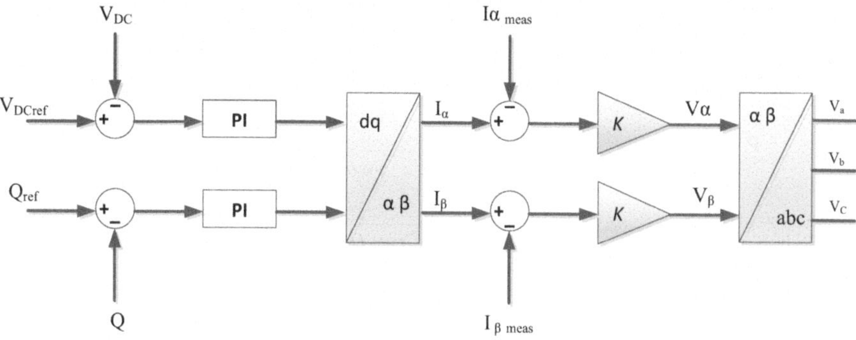

Fig. 5 Three-phase grid coupling inverter PQ control

However, in case of three-phase controllable MS, an active power set-point will define the MS power output. The three-phase DC-AC inverter PQ control is performed in a d-q reference frame, as represented in Fig. 5. The active power exported to the grid is controlled by the direct component of the reference current, generated by the DC voltage error. The reactive power output will be controlled by the quadrature component of the reference current resultant from the error between the measured value and the reference reactive power set-point. The d-q reference currents are then transformed to the α-β stationary reference frame. Afterwards, the inner current control loop based on a proportional controller is used in order to generate the converter output voltages [1].

Storage units are represented as ideal DC voltage sources coupled to the grid through an AC/DC inverter VSI. The DC/AC grid coupling inverter block diagram was described in Sect. 4 and is represented in Fig. 3.

5.1 EV Dynamic Model

The EV dynamic model is represented in Fig. 6. The EV grid connection was assumed to be single-phase (between a phase and the neutral conductor). Single-phase EV charging modes will result in longer charging times, offering higher flexibility and control possibilities. The DC-DC converter, linking the battery to the capacitor, takes the role of controlling the (dis)charging current on the battery side. On the other side, the AC-DC converter assures the link between the grid and the DC-link capacitor. The capacitor voltage is controlled to be constant by an adequate adjustment of the power flow with the AC grid. The variations seen in the DC-link capacitor voltage are due to the power demand/injection on the battery side. Therefore, the general control approach that is followed in the operation of the DC/AC EV power electronic interface is the same presented for single-phase PQ controlled inverters. The main differences rely in the external control loop which determines the EV charging power based on P-ω droop characteristic presented in Sect. 3. The EV charging power is determined based on the local frequency measurement.

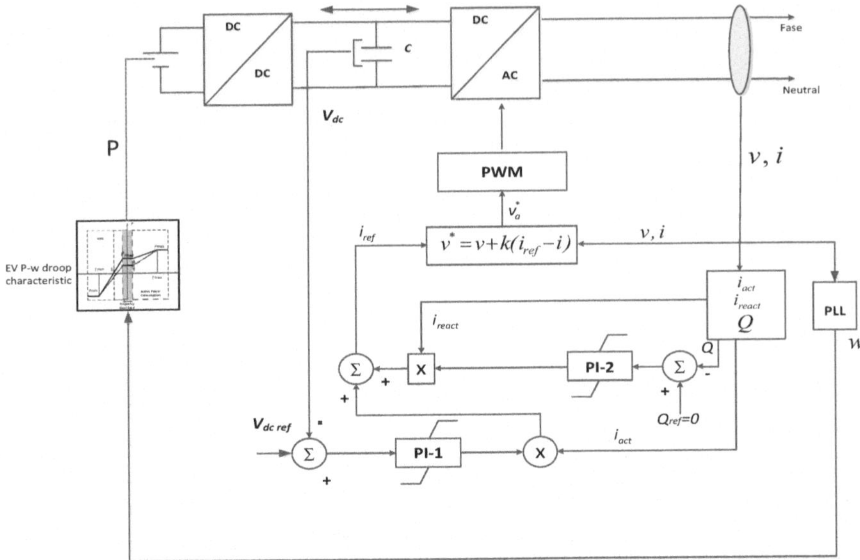

Fig. 6 Block diagram of EV dynamic model

The models developed were integrated in a simulation platform developed in a Matlab/Simulink environment through the use of the SimPowerSystems toolbox. The simulation platform was developed in order to analyze the dynamic behavior of the overall MG, together with the control strategies discussed in Sects. 3 and 4.

6 Evaluation of MG Stability and Voltage Quality During Islanded Operation

Ensuring successful MG autonomous operation can be quite challenging. In previous sections, innovative control strategies taking advantage of EV charging flexibility were proposed in order to improve the MG resilience and frequency regulation. Also, new voltage balancing strategies as well as the adoption of dynamic models adequate to the representation of the LV network as a three-phase four-wire systems were discussed. The main objective of this section is to illustrate the control strategies described before and to identify the main advantages resulting from the adoption of EV frequency-droop characteristics and from voltage balancing mechanisms.

6.1 MG Simulation Scenarios

For the development of the study presented in this section, two typical LV networks were considered: an urban and a rural LV distribution networks. The urban LV net-

Table 1 Urban MG scenario

	Type of connection					
	3~	A	B	C	Total	Deferrable loads
Load (kW)	38.5	69.8	32.4	37.1	177.8	104.65
MS (kW)	150	0	0	0	90	–
E.V. (kW)	0	22.3	24.3	6.73	57.7	–

Table 2 Rural MG scenario

	Type of connection					
	3~	A	B	C	Total	Deferrable loads
Load (kW)	87,9	44,4	38,3	46,6	217,2	175,2
MS (kW)	90	25,5	31,7	38,5	185,7	–
E.V. (kW)	0	21,9	19,6	17,6	59,8	–

Table 3 EV frequency-droop parameters

Frequency-droop parameters	Values (Hz)
Nominal frequency	50
Zero-crossing frequency	49.5
Maximum frequency	51
Minimum frequency	49
Frequency dead-band	0.2

work has a nominal voltage of 380 V and is composed by eight nodes with a total load of 177.8 kW, including three-phase and single-phase loads. The urban LV grid is built on underground cables with a total length of 312 m. The rural LV network is constituted by 13 nodes and has a total load of 217.2 kW. The rural LV grid is built on underground cables with a total length of 94 m and on overhead lines with a total length of 1,135 m. In order to build the MG simulation scenarios, a 150 kW storage unit was considered, to be placed at the MV/LV substation of both networks. In the urban scenario only three-phase power generating units were considered, namely single-shaft gas microturbines (SSMT). In the rural area network, a large amount of single-phase renewable-based microgeneration units was considered, namely wind turbines (WT) and solar photovoltaic panels (PV). The main characteristics of the rural and urban MG are summarized in Table 1 and 2 (the deferrable loads are those considered as candidates to be shed in case of severe frequency deviations following MG islanding).

As referred before, the EV are assumed to be connected to the MG through single-phase chargers. The maximum unitary EV charging power was considered to be 3 kW. At the nominal frequency the power absorbed was considered 75 % of the nominal power. Different EV frequency droop characteristics were obtained as a consequence of the EVs different charging rates. The EV frequency-droop parameters are presented in Table 3.

The MG simulation scenarios were implemented in the dynamic simulation platform described in Sect. 5.

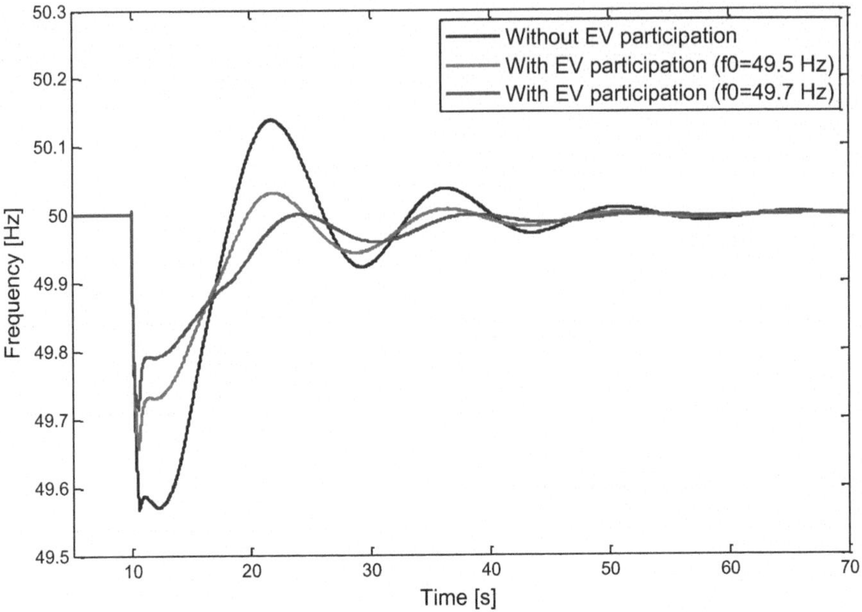

Fig. 7 Urban MG frequency response after the islanding

6.2 *MG Islanding with EV*

To study the benefits of the EV participation in frequency regulation, three cases were considered: (1) a base case where the EV do not respond to frequency variations, being regarded as conventional loads; (2) a case where the EV chargers are controlled with a frequency-droop strategy with a zero-crossing frequency of 49.5 Hz and (3) a case where the EV chargers are controlled with a frequency-droop strategy with a zero-crossing frequency of 49.7 Hz.

After 10 s, the MG is suddenly disconnected from the main grid and becomes isolated. In the moments subsequent to MG islanding, since the MG was importing a significant amount of power from the MV network, the system suffers a frequency drop, as represented in Fig. 7. In order to sustain the initial frequency drop, about 60 % of the MG total load is gradually disconnected. Secondary frequency control was implemented locally at each SSMT through PI controllers. After the islanding the secondary frequency control starts to respond immediately leading to the correction of the frequency deviation (the slow response of the SSMT leads to a significant time for frequency recovery).

In Fig. 7 it can also be observed that the EV participation significantly reduces the initial frequency deviation. Regarding MG frequency behavior, it is possible to observe that in Case 1 the frequency returns to the nominal value faster than in the base case. In fact, without the EV participation in the frequency regulation, only

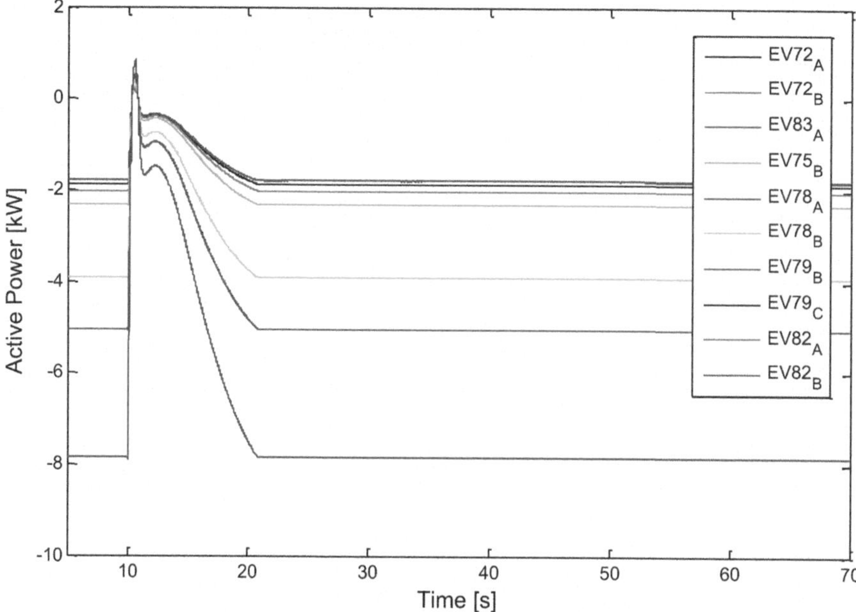

Fig. 8 Power absorbed/injected by the urban MG EV

the main storage unit and the SSMT (which has a slower response) are responsible for the frequency control, thus being responsible for a slower frequency restoration times.

Figure 8 presents the power absorbed/injected by the EV for Case 3. Increasing the zero-crossing frequency to 49.7 Hz forced the EV to inject power into the grid, consequently reducing the initial frequency deviation and allowing a faster recovery of the MG frequency to its nominal value. Consequently, the solicitation of the main storage device around the 10 s subsequent to the MG islanding is reduced, as represented in Fig. 9. In fact, in the moments subsequent to the islanding, the storage unit injects power to the grid to compensate the unbalance between load and generation. When the MG frequency reaches the nominal value (50 Hz) the power injected by the storage unit is zero.

6.3 *Voltage Balancing Mechanisms*

To study the importance of the voltage balancing unit, two cases were considered: a base case where the main storage unit is connected to the MG through a three-leg inverter with frequency and voltage droops; and a second case where the VSI is a four-leg inverter with the voltage balancing mechanism described in Sect. 4. In both cases, the EV participate in the MG frequency regulation. The benefits of

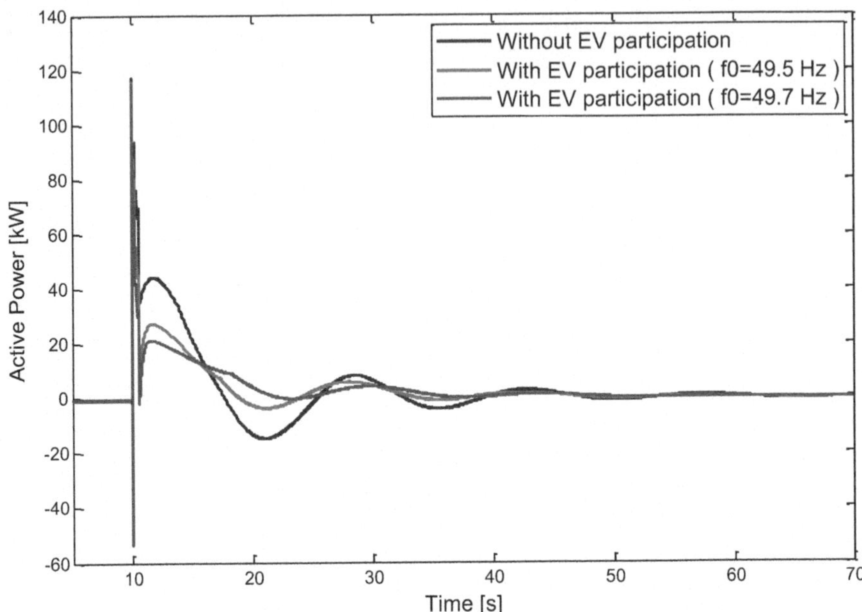

Fig. 9 Total power injected by the main storage unit with and without the EV participation in MG frequency regulation

the balancing unit will be quantified by the voltage unbalance factor (%VUF), also described in Sect. 4.

The MG simulation scenarios had considerable unbalances between the phases, due to an uneven distribution of single-phase loads and MS by the three phases of the system. However, the rural MG had higher voltage unbalance since it is constituted by long overhead lines with dispersed loads and has a considerable amount of single-phase MS. After the islanding there is 95.7 kW of single-phase power generation from the MS and 87.1 kW of both three-phase and single-phase loads.

Figures 10 and 11 present the %VUF at the VSI terminals with and without the balancing unit. Without the balancing unit both zero sequence and negative sequence voltage unbalance are higher than the 2 % limit imposed by the EN50160 standard [2]. However, the implementation of the voltage balancing mechanism eliminates both voltage components at the VSI output. This reduction will avoid large currents that would otherwise occur when the MG is synchronized and reconnected to the upstream network.

The effectiveness of the balancing unit depends on its distance to the unbalanced load/source. Since the Rural MG is constituted by long aerial lines, the voltage balancing effects are smaller at the nodes electrically more distant from the VSI. Figures 12 and 13 present the voltage unbalance at the node electrically more distant from the VSI, for the base case without the balancing unit and for the case with the balancing unit, respectively.

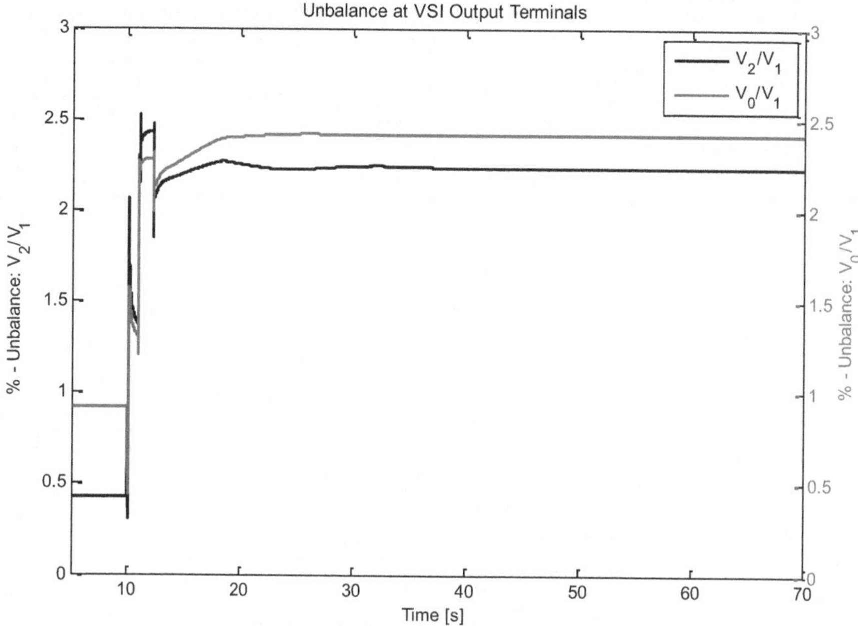

Fig. 10 VUF at the VSI terminals without the balancing unit

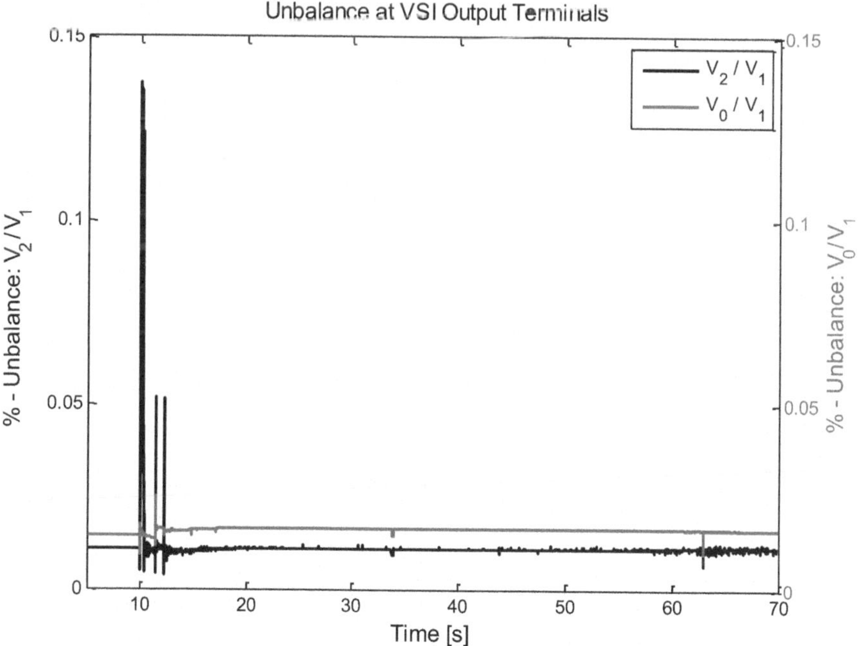

Fig. 11 VUF at the VSI terminals with the balancing unit

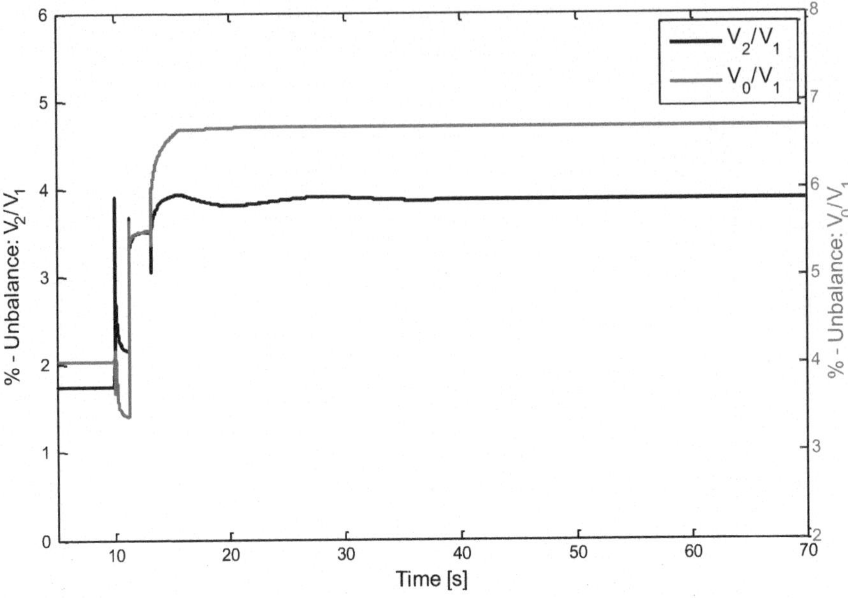

Fig. 12 Rural MG without the voltage balancing unit: VUF at the node electrically more distant from the VSI

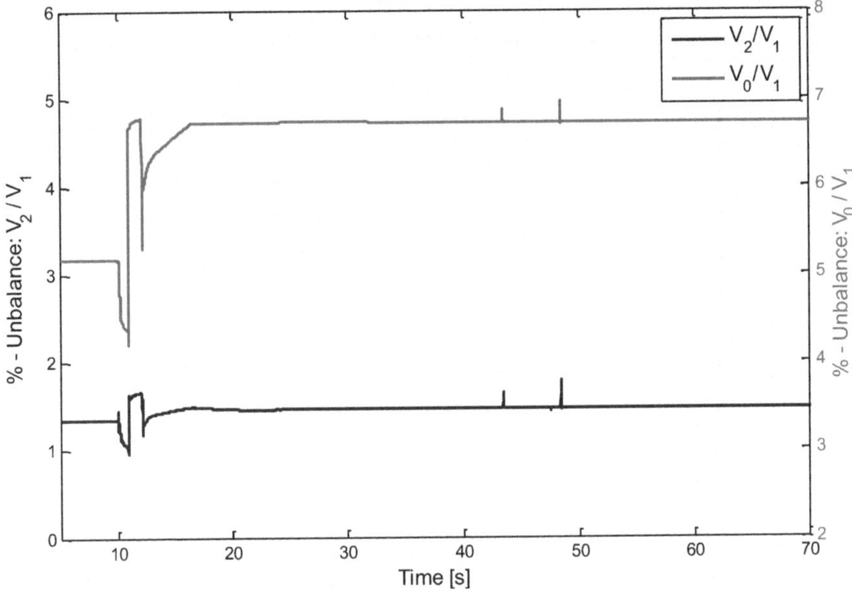

Fig. 13 Rural MG with the voltage balancing unit: VUF at the node electrically more distant from the VSI

In the base case, the voltage unbalance increases after the islanding, exceeding the acceptable limits. In fact, the single-phase MS such as PV and WT remain connected to the system regardless of the load shed, increasing the unbalance between the phases. When the balancing unit is considered, the negative sequence voltage unbalance is reduced to 1.5 % (V_2/V_1), being able to maintain the unbalance within the acceptable limits after the MG islanding. However, the balancing unit has no effect on the zero sequence voltage components, which remains at 6.8 % (V_0/V_1), after the islanding. The results obtained reinforce the need of developing local voltage regulation mechanisms for unbalanced LV networks.

7 Conclusions

Developing new distribution operating strategies to increase the system security and resilience to major power system disturbances is a key issue for the development of the smart grid vision. In this sense, the MG introduces a new distribution operating philosophy capable of managing the resources connected downstream, such as microgeneration, EV, storage, and responsive loads in order to operate autonomously.

The active participation of EV and MS in MG emergency operation can improve EV participation in the frequency regulation and leads to a smoother MG transition from the interconnected mode to unplanned islanding. The results obtained from numerical simulation demonstrate that the adoption of such scheme reduces the initial frequency deviation and leads to a faster recovery of the MG frequency to the nominal value. The EV frequency-droop strategy is flexible enough to reflect the EV chargers characteristics, EV owner willingness to participate in the MG operation, and to be coordinated with the other frequency and voltage regulation mechanisms. As a consequence of the immediate response of EV to the frequency deviation, it was demonstrated that the adopted EV frequency control strategy reduces the power solicitations to the MG main storage unit. However, the impact on the main storage state of charge will depend on the MG pre-fault load/generation conditions as well as on the number of EV connected to the network.

Regarding the MG voltage balancing mechanisms, the obtained results demonstrate the feasibility of the proposed approach for the cancelation of unwanted negative and zero sequence voltage components at the VSI terminals where it was installed. It was also possible to observe that in other MG nodes a significant reduction of voltage unbalance was obtained. However, the effectiveness of the balancing unit will depend on its distance to the unbalanced load or source, requiring the development and identification of additional voltage balancing strategies.

References

1. Blaabjerg F, Teodorescu R, Liserre M, Timbus AV (2006) Overview of control and grid synchronization for distributed power generation systems. IEEE Trans Ind Electron 4:1398–1409
2. CENELEC EN 50160 (1994) Voltage characteristics of electricity supplied by public distribution systems
3. Engler A, Burger B (2001) Fast signal conditioning in single phase systems. In: Proceedings 9th European Conference on Power Electronics and Applications, Graz, Germany, 27-29 August 2001
4. Guerrero JM, Vasquez JC, Matas J, De Vicuna LG, Castilla M (2011) Hierarchical control of droop-controlled AC and DC microgrids-a general approach toward standardization. IEEE Trans Ind Electron 1:158–172
5. Lasseter B (2001) Microgrids (distributed power generation). In: IEEE power engineering society winter meeting, vol 1, 28 Jan–1 Feb 2001, pp 146–149
6. Lasseter RH (2011) Smart distribution: coupled microgrids. In: Proceedings of the IEEE, vol 99, no 6, pp 1074–1082
7. Li Y, Vilathgamuwa DM, Loh PC (2005) Microgrid power quality enhancement using a three-phase four-wire grid-interfacing compensator. IEEE Trans Ind Appl 6:1707–1719
8. Moreira C (2008) Identification and development of microgrids emergency control procedures. PhD thesis, Faculdade de Engenharia da Universidade do Porto
9. Moslehi K, Kumar R (2010) A reliability perspective of the smart grid. IEEE Trans Smart Grid 1:57–64
10. Pecas Lopes JA, Moreira CL, Madureira AG (2006) Defining control strategies for microgrids islanded operation. IEEE Trans Power Syst 2:916–924
11. Pecas Lopes JA, Polenz SA, Moreira CL, Cherkaoui R (2010) Identification of control and management strategies for LV unbalanced microgrids with plugged-in electric vehicles. Electr Power Syst Res 8:898–906
12. Pecas Lopes JA, Soares FJ, Almeida PMR (2011) Integration of electric vehicles in the electric power system. In: Proceedings of the IEEE, vol 99, no 1, pp 168–183
13. Zmood DN, Holmes DG, Bode GH (2001) Frequency-domain analysis of three-phase linear current regulators. IEEE Trans Ind Appl 2:601–610

An Optimized Adaptive Protection Scheme for Distribution Systems Penetrated with Distributed Generators

Ahmed H. Osman, Mohamed S. Hassan and Mohamad Sulaiman

1 Introduction

Traditional distribution systems are designed to be radial in nature. As a result, their protection systems are designed and set based on one utility source feeding the whole system. Nowadays, distributed generators (DGs) are increasingly connected to the distribution systems to meet the load demand and increase the reliability of the system. With the additional connected sources, the system is no longer radial. Moreover, during a fault condition, the fault is being fed from all the sources connected to the power system. Therefore, the fault current level is expected to be different from the case of a radial system. The DGs will affect the protection relays of the distribution feeder in such a way that it will reduce the reach of the relay. This is due to the fact that the DG will increase the equivalent impedance of the feeder which will decrease the fault current. Furthermore, the protection relays in the main feeder can see fault currents in forward or reverse directions and they have to detect the fault direction [1–2].

One more important problem in distribution systems penetrated with DGs is the varying topology of the network whenever a DG is disconnected from the system for any reason. This problem arises due to the fact that one setting for the protection relays cannot respond correctly to the changing topology of the system and will affect the safety and reliability of the distribution system. Thus, to avoid relay malfunction, the relays have to be adaptively set whenever a new system topology is detected.

A. H. Osman (✉) · M. S. Hassan · M. Sulaiman
Electrical Engineering, American University of Sharjah, Sharjah, UAE
e-mail: aosmanahmed@aus.edu

M. S. Hassan
e-mail: mshassan@aus.edu

M. Sulaiman
e-mail: b00023328@aus.edu

R. Karki et al. (eds.), *Reliability Modeling and Analysis of Smart Power Systems,* Reliable and Sustainable Electric Power and Energy Systems Management,
DOI 10.1007/978-81-322-1798-5_4, © Springer India 2014

Adaptive protection is "an online activity that modifies the preferred protective response to a change in system conditions or requirements in a timely manner by means of externally generated signals or control action" [3]. Adaptive protection of distribution systems penetrated with distributed generation can be realized with the use of microprocessor-based relays that have the advantage of easily changing their tripping characteristics. Implementing adaptive protection increases the sensitivity and achieves faster operating times.

Adaptive protection can be achieved by measuring the local voltage and frequency to detect the presence of the DGs when they are connected or disconnected to the power system as given in [4]. When a fault is detected and isolated by certain relays, the settings of the other relays will be changed. To further reduce the operation times, communication-based approaches could be integrated to update the settings of the protection relays.

A multiagent approach to power system protection coordination has been proposed in [5]. The coordination strategy is embedded in every relay agent to coordinate them under different system situations. The feasibility of the multiagent system has been demonstrated by communication simulation. A cooperative protection system with a multiagent system is proposed in [6]. The system allows the cooperation between a number of equipment and various adaptive protection functions can be realized to fit the changes in the power system automatically. A multiagent-based adaptive wide-area current differential protection system in which an expert system is integrated within the SCADA system to divide the power grid into different protection zones is proposed in [7]. A multiagent technology into relay protection setting for power plant is introduced in [8]. The multiagent system connects agents that can work independently, cooperate, and share knowledge.

In this chapter, a novel optimally coordinated adaptive overcurrent protection relaying strategy is proposed. The proposed scheme utilizes communication links between the digital overcurrent relays and a central relaying unit (CRU). Any change in the connection status of the distributed generators initiates the adaptive coordination strategy. Also, the scheme can identify the faulty section of the line to speed up the isolation time.

2 Optimization of Relay Coordination

Traditionally, the coordination can be achieved by topology [9, 10], optimization [11, 12], and intelligent methods [13]. Topological analysis is employed for setting relays in multi-loop networks. Graph theoretic and functional dependency approaches are applied to provide a solution which is the best alternative setting, but not necessarily an optimal solution. In optimization-based methods, nonlinear programming is commonly used to determine the optimal time dial settings of the relays subjected to the coordination constraints as well as the limits of the relay setting [11, 14]. Linear programming techniques are applied to minimize the operating time, while the pickup current is selected based on experience or by using a generalized reduced gradient technique [12]. Intelligence methods are also used in searching the optimized setting for relay coordination [5, 13].

In the optimization technique, the problem is typically stated as an optimization problem with a main objective function that is subjected to coordination constraints, relay characteristics, and the limits of the relay settings [15]. Optimal coordination needs to be performed to maintain all the directional relays in the system properly coordinated and ready for fault occurrence. The optimal coordination will select two factors which are the TDS and the tap setting (pickup current). The selection of those two fundamental settings should satisfy the requirements of sensitivity, selectivity, reliability, and speed. Thus, the TDS values and tap setting of each relay should minimize the overall operation time of the relays, while maintaining the requirements intact.

There is more than one approach to implement optimal coordination. First, the pickup current can be fixed and linear programming (LP) can be used to solve the optimization problem. The pickup current can also be set to be continuous and non-linear programming (NLP) can be used. The last approach is by having the pickup current to be discrete values where mixed-integer nonlinear programming (MINLP) can be used [16].

Since the nonlinearity is introduced by the pickup currents, the pickup current for a relay can be written as a sum of available pickup currents for the relay. Each pickup current is multiplied by a binary variable which decides which pickup current is chosen for the relay. The pickup current formulation is given by:

$$I_{pi} = \sum_{m} y_{mi} I_{pam}. \tag{1}$$

In (1), y_{mi} will be equal to "1" if setting m is chosen while it is "0" otherwise [17]. The objective of all those approaches is to minimize the overall operation time of the relays. This means that in all the approaches the TDSs are going to be initially set to the minimum and then gradually start increasing its value.

To solve the optimization problem, an objective function needs to be formulated along with the constraints that will help in achieving better coordination. According to [16], the objective function is the sum of all the relay operation times of the relays and is given by:

$$objective = \min \sum W_i' T_{ik} \tag{2}$$

where W_i' represents the weights which are usually set to 1 and T_{ik} is the operation time of relay R_i for a fault in zone k. Moreover, the formula to calculate the operation time of the relays T_{ik} is given by:

$$T_{ik} = \frac{0.14\, TDS_i}{\left(\frac{I_{ik}}{I_{pi}}\right)^{0.04} - 1} \tag{3}$$

where I_{ik} is the fault current passing through the relay. If the pickup current I_{pi} is unknown, Eq. (3) becomes nonlinear. In LP the pickup current is assumed to be known and as a result Eq. (3) will be transformed to:

$$T_{nk} = \alpha_i \times TDS_i \tag{4}$$

where α_i is a constant value that varies from one relay to the other depending on the values of the selected pickup current [16].

Constraints are added to ensure that the relays are properly coordinated. Those constraints relate the operation times of the primary and backup relays, and to maintain a minimum coordination time between the relays. This is done to prevent a backup relay from taking an action before the operation of the primary relay. Eq. (5) shows the constraint equation. In this equation, T_{nk} is the operation time of the first backup relay R_i in zone k, and ΔT is the coordination time between the primary and backup relays.

$$T_{nk} - T_{ik} \geq \Delta T \tag{5}$$

To complete the optimization problem, boundaries have to be set to limit the values of TDS and pickup current. Those boundaries are shown in Eqs. (6) and (7). According to [16], the pickup current can be from 50 to 200 % in steps of 1 %, while the TDS can be set from 0.01 to 1 in steps of 0.01.

$$TDS_{i\min} \leq TDS_i \leq TDS_{i\max} \tag{6}$$

$$I_{pi\min} \leq I_{pi} \leq I_{pi\max}. \tag{7}$$

The pickup current limits are chosen so that the lower limit is equal to maximum load current after being multiplied by 1.3, and the upper limit is equal to the minimum fault current at the far end of the faulted bus [18].

3 Relay Communication

Online, speedy, and efficient information exchange between relays and the CRU can be enabled by integrating communication-based approaches with adaptive protection systems. The distributed network protocol (DNP) and the IEC 61850 are two widely accepted and commonly used protocols to support communications in industrial and power systems applications.

The DNP3 was first presented in 1993 by GE, and it is based on the IEC 60870-5 protocol [19]. It was created to serve in supervisory control and data acquisition (SCADA) applications [19]. Because of its well-known reliability, efficiency, robustness, and wide compatibility [20], DNP3 is recommended as the communication

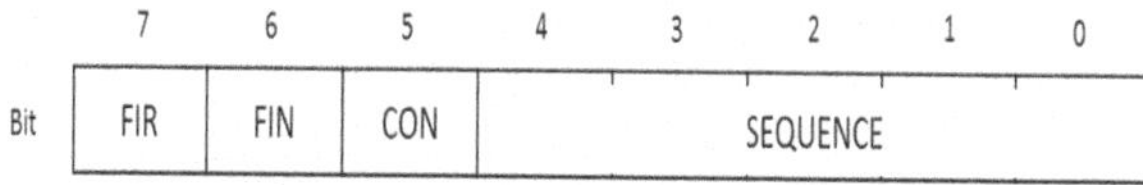

FIR	First fragment of multi-fragment message
FIN	Final fragment of multi-fragment message
CON	Confirmation required
Sequence	Fragment sequence number 0 – 15 Master requests and outstation responds (solicited) 16 – 31 outstation unsolicited responses

Fig. 1 Application control byte [19]

protocol in the proposed adaptive protection system. This will allow easier system integration between the proposed technique and the existing equipment. DNP3 communicates through messages. Two types of messages are exchanged during DNP3 operation. In the proposed scheme, the first message type carries the connection status of the DGs while the second message type carries the TDS and tap setting values that are sent to the relays to change their existing settings upon any change in the system topology.

The message buildup process in DNP3 works as follows. First, the application layer formats the request and response messages. In addition, it initially segments the data into manageable sized blocks that are called application service data units (ASDU). Then, the application protocol control information (APCI) is added, which in turn forms the application protocol data units (APDU), each of which has a maximum size of 2048 bytes [20]. The size of an APCI message is either 2 or 4 bytes depending on whether the message is a request or a response. When there is only a command and no user data, the message does not contain an ASDU. It consists only of a header which is added to the user data, and contains the application control byte, function code, and internal indications. The application control byte indicates if the message is part of a multi fragment, and whether a confirmation is requested or not. In addition, it indicates if the fragment is unsolicited, and provides the sequence number that allows for the detection of dropped or lost messages. Figure 1 shows the function of the application control byte.

After that, the APDU is passed to the transport layer which breaks down the data into 249-byte fragments called the transport protocol data units (TPDU). Then, a one-byte transport header is added resulting in TPDU of a total size of 250 bytes. This header byte contains information indicating whether it is the first or last fragment in addition to the sequence number. After that, the TPDU is passed to the data link layer.

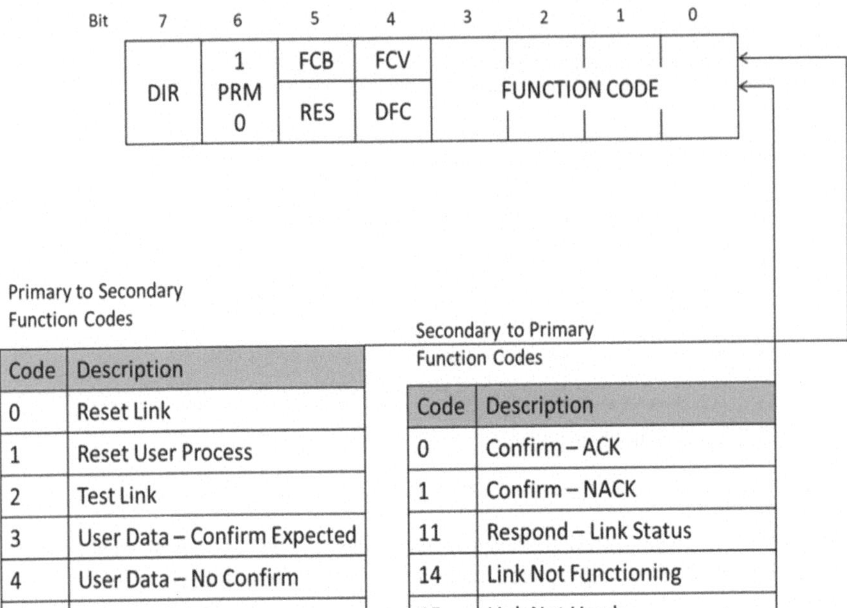

Code	Description
0	Reset Link
1	Reset User Process
2	Test Link
3	User Data – Confirm Expected
4	User Data – No Confirm
9	Request Link Status

Code	Description
0	Confirm – ACK
1	Confirm – NACK
11	Respond – Link Status
14	Link Not Functioning
15	Link Not Used

Fig. 2 Control byte [19]

The data link layer then adds a 10-byte header in addition to a 32-byte cyclic redundancy check (CRC) code to the user data. The link protocol data unit (LPDU) has a total of 292 bytes [20]. The added data link header contains a start, length, control, destination address, source address, and a 16-bit CRC. The control byte provides the control of data flow over the physical link, identifies the data types, and indicates its direction. The bit functions of the control byte are shown in Fig. 2.

For the physical channel, optical fiber links can be used as the physical medium for communication between the relays and the CRU. Optical fiber is typically recommended as a communication medium because of its reliability and immunity to noise.

In the simulations of the proposed adaptive scheme, a 1 Gbps optical fiber channel is assumed. Furthermore, the channel is assumed to be error-free and hence only the delay of the physical medium is taken into account. The simulation results indicated a delay of 5 ns when receiving the connection status of the DGs while the experienced delay when transmitting the TDS and tap settings to the relays was 1.152 μs. The proposed communication topology between the DGs, relays, and the CRU is depicted in Fig. 3.

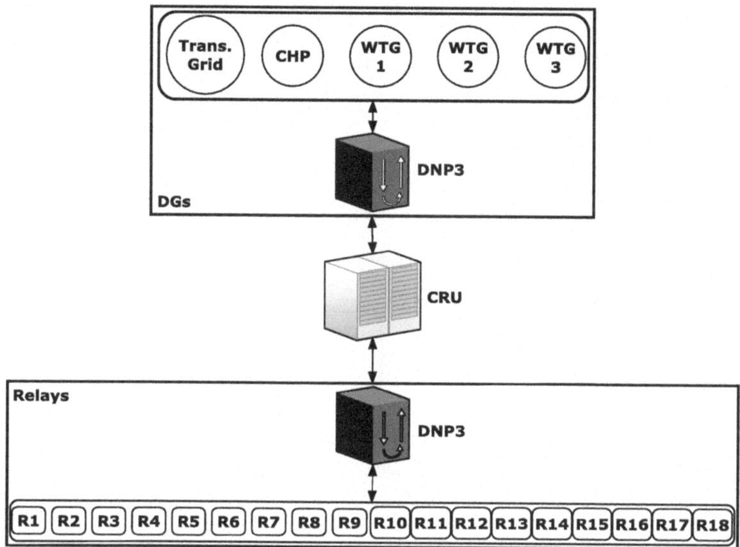

Fig. 3 Proposed communication topology

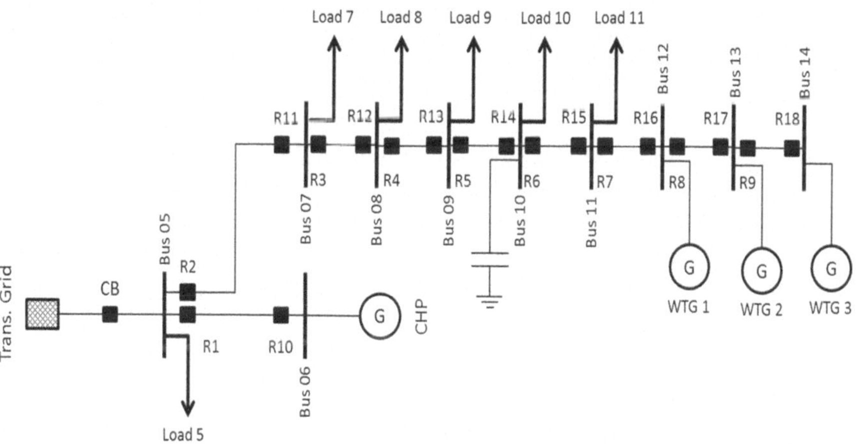

Fig. 4 System Model

4 System Model

To test the proposed adaptive protection strategy, the power system studied in [21] is adopted. The system is given in Fig. 4 and its data is given in Appendix. It consists of 10 buses connecting nine line sections, three fixed wind turbine generators (WTGs), one combined heat and power (CHP) plant, and six loads.

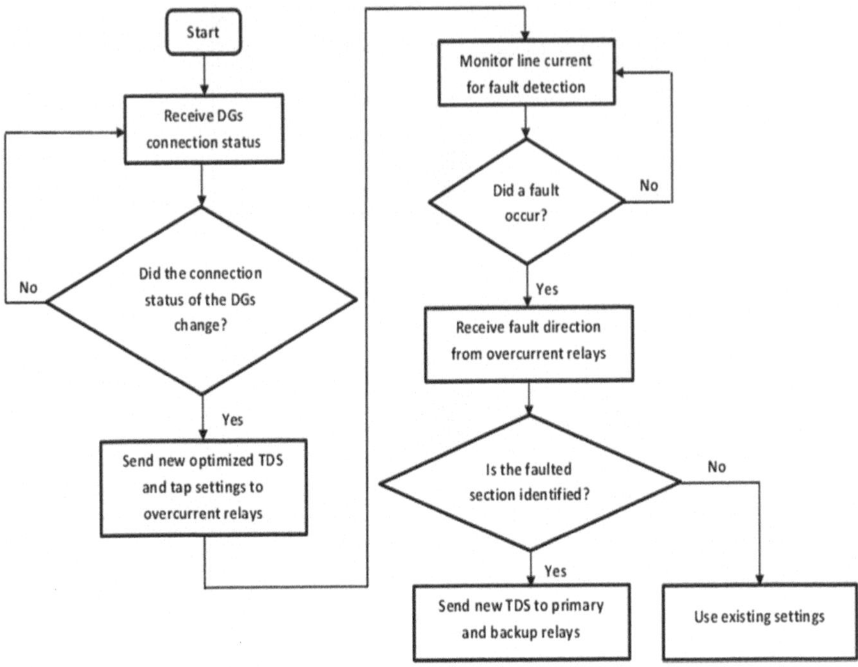

Fig. 5 Flow chart of proposed methodology

The system is modeled using PSCAD. Eighteen directional overcurrent relays are used in the modeled system, where each line section is protected by two directional overcurrent relays. Fault analyses at different fault locations, different fault types, and different connection status of the DGs were performed using PSCAD. This data was stored in databases which are used to test the proposed adaptive protection system strategy.

5 Proposed Adaptive Protection Strategy

The proposed adaptive protection strategy is shown in Fig. 5. The generators' connection statuses are received at a CRU using a fiber optic communication channel which utilizes the DNP3 protocol. The received signal is either "1" in case a generator is connected or "0" when a generator is disconnected.

After determining the connection status of the DGs, the database is used to provide the fixed maximum load currents, maximum fault currents, and minimum fault currents for the current system configuration. The CT ratios, which are fixed, are selected using the maximum load currents. The CT ratios are selected using 125 % of the maximum load current at each relay. The tap settings are changed based on the system configuration and it is selected using the calculated minimum fault current at each relay.

Linear optimization is implemented to determine the optimized values of TDS for each relay. A moderately inverse-time overcurrent relay characteristic curve is used in the optimization process. The objective function, given in Eq. (2), is formulated, and optimized based on the constraint Eqs. (5), (6), and (7). The values of the tap settings which were previously determined are used to calculate the coefficients of the objective equation. The optimized TDS values along with the calculated tap settings for the current configuration are communicated to the relays to update their settings and be ready for any fault condition.

It is known that overcurrent relaying protection suffers from the delay in the fault clearance especially if the fault is close to the source and due to the relay coordination requirements. A strategy to overcome this problem using the available communication links between the directional overcurrent relays and the CRU is proposed.

The strategy to speed up the fault clearance is based on the identification of the faulted section of the line using the directional element of the relays. The fault direction is identified using a negative sequence directional element at each relay [22]. Upon the occurrence of a fault, each of the directional overcurrent relays sends the direction of the fault that it sees to the CRU. The faulted section is identified when each of the two relays at the beginning and end of a line section see the fault in the forward direction. Once the faulted section is identified, the CRU sends new TDS and tap settings to the relays. The new settings are part of a database stored in the CRU for different possibilities of faulted sections in the system under test. These new settings ensure that the relays in the faulted section have the fastest response followed by a slower response for the backup relays on both sides. If for any reason the directional elements fail to identify the faulted line section, the fault will be cleared based on the settings that were previously given to the relays.

6 Results

To test the operation of the proposed adaptive protection strategy, the system configuration was first changed to observe its impact on the optimized TDS and tap setting selections for the relays. The system is simulated when all the DGs are connected using PSCAD. The results of the optimized settings for the 18 relays are shown in Table 1.

With the CHP plant disconnected from the system, the algorithm at the CRU should run again to determine the optimized TDS and tap settings for the new system configuration. The new TDS and tap settings are presented in Table 2. Similar setting results can be obtained for any other DG disconnection or any system configuration.

The proposed communication-based adaptive scheme is tested during fault conditions to check the coordination of the relays and to evaluate the effect of identifying the faulted section on the operating time. In this case, the considered system

Table 1 TDS and tap settings without DG disconnection

Relay	Tap setting	TDS
1	4.9	0.2494
2	0.5	3.8172
3	2.4	3.0920
4	2.0	2.6085
5	1.9	2.1294
6	2.1	1.5741
7	4.0	1.0191
8	2.7	0.6228
9	1.8	0.1307
10	4.9	0.1471
11	3.2	0.1146
12	2.9	0.4713
13	2.6	0.9086
14	2.2	1.4129
15	3.2	1.8697
16	4.0	2.2438
17	3.6	2.7216
18	2.7	3.2696

Table 2 TDS and tap settings after CHP disconnection

Relay	Tap setting	TDS
1	0.1	0.3469
2	0.5	3.8143
3	2.4	3.0897
4	2.0	2.6068
5	1.9	2.1280
6	2.1	1.5731
7	4.0	1.0184
8	2.7	0.6224
9	1.8	0.1307
10	10.0	9.7920
11	3.2	0.1146
12	2.9	0.4709
13	2.6	0.9078
14	2.2	1.4118
15	3.2	1.8682
16	4.0	2.2420
17	3.6	2.7194
18	2.7	3.2699

configuration is when all the generators are connected. A single-line-to-ground (SLG) fault on the line section between buses 11 and 12 is simulated. The directional overcurrent relays in the feeder sections enabled the CRU to identify the faulted section as previously explained in Sect. 5. Then, the CRU will send new optimized settings to the relays from the stored database. For the sake of comparison, the

Table 3 TDS, tap settings, and operation times for a SLG fault between buses 11–12 (All generators connected)

		Faulted section is not identified		Faulted section is not identified	
Relay	Tap setting	TDS	Operation time (s)	TDS	Operation time (s)
1	4.9	0.2494	Blocked	0.2494	Blocked
2	0.5	3.8172	2.8723	2.9247	2.2007
3	2.4	3.0920	2.2635	2.2558	1.6514
4	2.0	2.6085	1.7645	1.7392	1.1765
5	1.9	2.1294	1.3604	1.2204	0.7797
6	2.1	1.5741	0.9955	0.6533	0.4132
7	4.0	1.0191	0.6739	0.1319	0.0872
8	2.7	0.6228	Blocked	5.5500	Blocked
9	1.8	0.1307	Blocked	5.5500	Blocked
10	4.9	0.1471	No operation	0.5049	No operation
11	3.2	0.1146	Blocked	5.5500	Blocked
12	2.9	0.4713	Blocked	5.5500	Blocked
13	2.6	0.9086	Blocked	5.5500	Blocked
14	2.2	1.4129	Blocked	5.5500	Blocked
15	3.2	1.8697	Blocked	5.5500	Blocked
16	4.0	2.2438	1.6276	0.1532	0.1111
17	3.6	2.7216	1.9860	0.6565	0.4791
18	2.7	3.2696	2.3995	1.2304	0.9030

settings of the relays and their operating times with and without the application of faulted section identification are shown in Table 3.

It can be seen that the faulted section identification scheme reduced the operation times of the primary relays R_7 and R_{16} from 0.6739 s and 1.6276 s to 0.0872 s and 0.1111 s, respectively. Furthermore, the two backup relays, R_6 and R_5, for the primary downstream looking relay, R_7, with respect to the transmission grid (TG) are coordinated. Similarly, the two backup relays, R_{17} and R_{18}, for the primary upstream looking relay, $R_{13,}$ with respect to the TG are also coordinated. The minimum coordination time used is 0.3 s. In Table 3, "No operation" indicates that the current seen by the relay is less than the pickup current, while "Blocked" indicates that the current is above the pickup current but the relay sees a reverse fault. Figure 6a, b show the time-current characteristics of the primary and the two backup relays in both directions with and without faulted section identification.

The performance of the proposed scheme is also tested during islanding operation mode. While the TG is disconnected and the system is allowed to operate in the islanding mode, a SLG fault has occurred on the line between buses 11 and 12. Table 4 only shows the settings and operation times of the primary and the two backup relays in both directions. The omitted relays are either not operating or blocked. Table 4 also shows the settings of the relays and the operating times for the two cases with and without the application of faulted section identification. The faulted section identification scheme reduced the operation times of the primary relays R_7 and R_{16} from 0.6422 s and 1.5101 s to 0.0948 s and 0.1121 s, respectively. Figure 7a, b show the time-current characteristics of the primary and the two backup relays in both directions with and without faulted section identification.

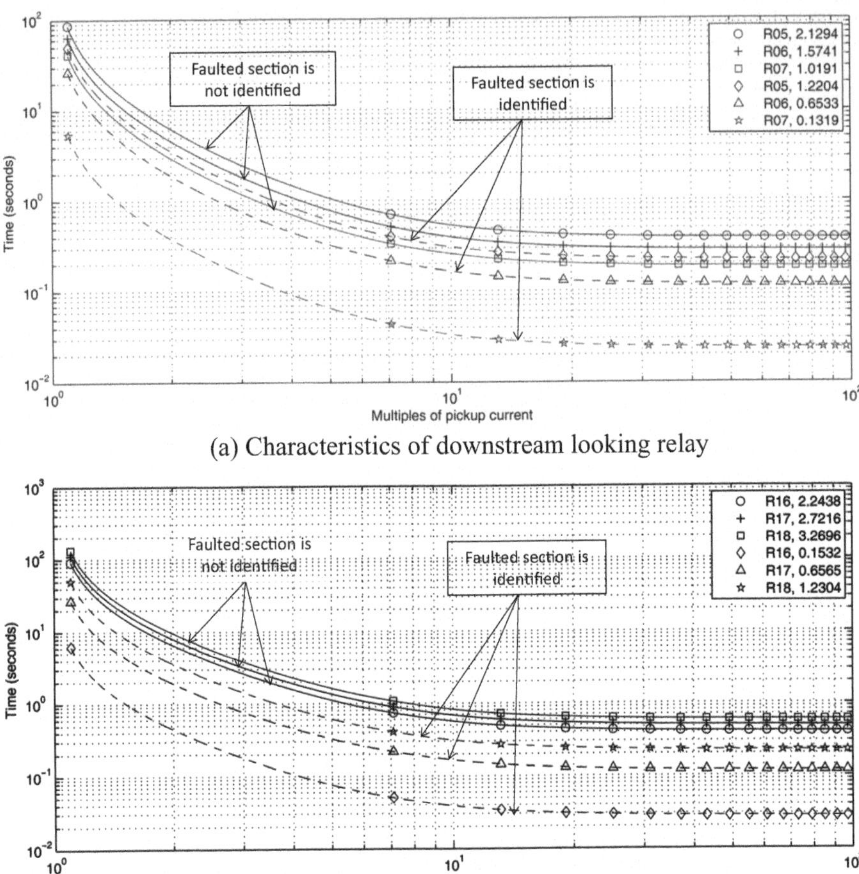

Fig. 6 Upstream and downstream TDS curves for a SLG fault on the line between buses 11 and 12 with all generators connected. **a** Characteristics of downstream looking relay. **b** Characteristics of upstream looking relays

Table 4 TDS, tap settings, and operation times for a SLG fault between buses 11–12 (islanded operation)

		Faulted section is not identified		Faulted section is not identified	
Relay	Tap setting	TDS	Operation time (s)	TDS	Operation time (s)
5	1.6	1.6668	1.2589	0.9995	0.7549
6	2.0	1.2770	0.9423	0.5629	0.4154
7	3.8	0.8245	0.6422	0.1217	0.0948
16	3.8	2.0522	1.5101	0.1523	0.1121
17	3.4	2.4562	1.8148	0.6488	0.4794
18	2.6	2.8446	2.1212	1.2100	0.9023

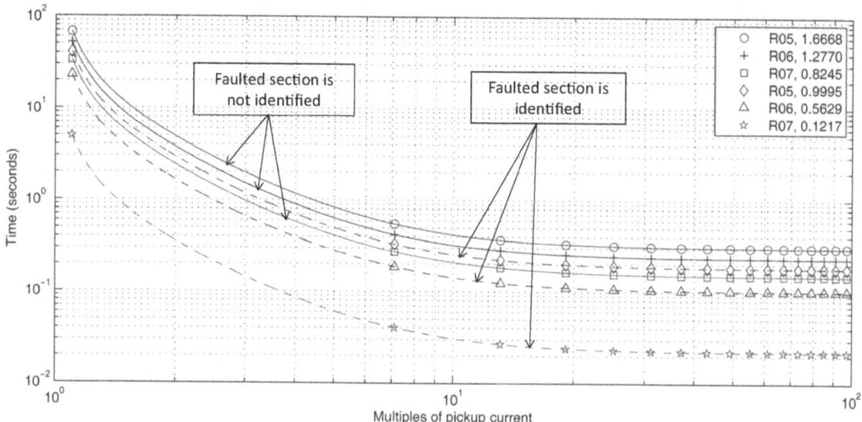

(a) Characteristics of downstream looking relays

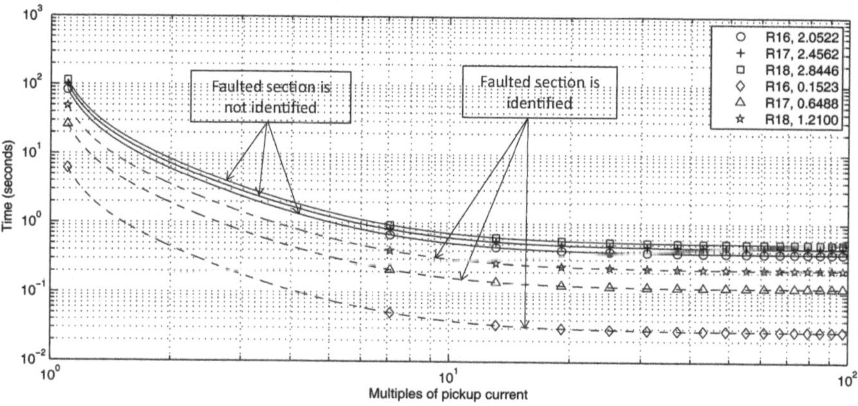

(b) Characteristics of upstream looking relays'

Fig. 7 Upstream and downstream TDS curves for a SLG fault on the line between buses 11 and 12 during islanding operation. **a** Characteristics of downstream looking relays **b** Characteristics of upstream looking relays'

7 Conclusions

An optimized communication-based adaptive protection scheme for distribution systems penetrated with DGs is proposed and tested. The integrated communication-based system ensured fast, efficient, and optimized fault clearance. The optimized settings of the relays are automatically updated whenever there is a change in the connection status of the connected DGs to the system. Also, the operating time of the relays is greatly reduced when the faulted line section is identified and new optimized settings are sent to the relays.

Appendix

Line data for the modeled system

From bus	To bus	Resistance (Ω)	Reactance (Ω)
5	6	0.1256	0.1404
5	7	0.1344	0.0632
7	8	0.1912	0.0897
8	9	0.4874	0.2284
9	10	0.1346	0.0906
10	11	1.4555	1.1130
11	12	0.6545	0.1634
12	13	0.0724	0.0181
13	14	0.7312	0.3114

Generators data

Parameters	CHP	WTG
Type of generator	Synchronous	Asynchronous
Number of parallel machine	3	1
Transformer to connect to grid	3.3 MVA, 20/6.3 kV	630 kVA, 20/0.4 kV
Individual generator's rating		
Rated power	3.3 MW	630 kW
Rated voltage	6.3 kV	0.4 kV
Stator resistance	0.0504 p.u.	0.018 p.u.
Stator reactance	0.1 p.u.	0.015 p.u.
Synchronous reactance d-axis	1.5 p.u.	–
Synchronous reactance q-axis	0.75 p.u.	–
Transient reactance d-axis	0.256 p.u.	–
Sub-transient reactance d-axis	0.168 p.u.	–
Sub-transient reactance q-axis	0.184 p.u.	–
Transient time constant d-axis	0.53 s	–
Sub-transient time constant d-axis	0.03 s	–
Sub-transient time constant q-axis	0.03 s	–
Mag. reactance	–	4.42 p.u.
Rotor resistance	–	0.0108 p.u.
Rotor reactance	–	0.128 pu.
Inertia time constant	0.54 s	0.38 s

Transmission grid (TG) data

Parameters	Value
Maximum short circuit power	10,000 MVA
Minimum short circuit power	8,000 MVA
Maximum R/X ratio	0.1
Maximum Z2/Z1 ratio	1
Maximum X0/X1 ratio	1
Maximum R0/X0 ratio	0.1

Load and generation data

Bus	PG (MW)	QG (Mvar)	PL (MW)	QL (Mvar)
05	0	0	3.87	0.85
06	6	0	0.0	0.0
07	0	0	0.56	0.11
08	0	0	0.56	0.11
09	0	0	0.55	0.10
10	0	1.5	0.85	0.20
11	0	0	0.51	0.13
12	0.31	0	0.0	0.0
13	0.31	0	0.0	0.0
14	0.31	0	0.0	0.0

References

1. Kaufmann M (1995) Power system protection: principles and components, vol. 1, T E T Association (ed). University Press, Cambridge, p 525
2. Baran M, El-Markabi I (2004) Adaptive over current protection for distribution feeders with distributed generators. In: Power systems conference and exposition, IEEE PES
3. Rockefeller G, Wagner C, Linders J, Hicks K, Rizy D (1988) Adaptive transmission relaying concepts for improved performance. IEEE Trans Power Del 3(4):1446–1458
4. Mahat P, Chen Z, Bak-Jensen B, Bak CL (2011) A simple adaptive over-current protection of distribution systems with distributed generation. IEEE Trans Smart Grid 2(3):1–10
5. Wan H, Li K, Wong K (2010) An adaptive multiagent approach to protection relay coordination with distributed generators in industrial power distribution system. IEEE Trans Ind Appl 46(5).2118–2124
6. Yang M-Y, Zhu Y-L (2005) Study on adaptive distance protection using multi-agent technology. In: The 7th International Power Engineering Conference. IPEC 2005.
7. Sheng S, Li K, Chan W, Zeng X, Shi D, Duan X (2010) Adaptive agent-based wide-area current differential protection system,. IEEE Trans Ind Appl 46(5):2111–2117
8. Zhao Q, Liu S (2011) Relay protection based on multi-agent system. In: 2011 International Conference on Electrical and Control Engineering (ICECE), Yichang, 16–18 Sept 2011
9. Knable AH (1969) A standardized approach to relay coordination, In: IEEE Winter Power Meeting, New York
10. Jenkins L, Khicha H, Shivakumar S, Dash P (1992) An application of function dependencies to the topological analysis of protection schemes. IEEE Trans Power Del 7(1):77–83
11. Urdaneta AJ, Nadira R, Jiménez LGP (1988) Optimal coordination of directional overcurrent relays in interconnected power systems. IEEE Trans Power Del 3(3):903–911
12. Chattopadhyay B, Sachdev MS, Sidhu TS (1996) An on-line relay coordination algorithm for adaptive protection using linear programming technique. IEEE Trans Power Del 11(1):165–173
13. So CW, Li KK (2000) Time coordination method for power system protection by evolutionary algorithm. IEEE Trans Ind Appl 36(5):1235-1240
14. Laway NA, Gupta HO (1993) A method for adaptive coordination of overcurrent relays in an interconnected power system. In: Fifth International Conference on Developments in Power System Protection
15. Urdaneta A, Perez L, Restrepo H (1997) Optimal coordination of directional overcurrent relays considering dynamic changes in the network topology. IEEE Trans Power Del 12(4):1458–1464

16. Zeineldin H, El-Saadany E, Salama M (2005) Optimal coordination of directional overcurrent relay coordination. In: Power Engineering Society General Meeting, 2005, IEEE. Vol 2, p 12, June 2005
17. Zeienldin H, El-Saadany E, Salama M (2004) A novel problem formulation for directional overcurrent relay coordination. In: Large Engineering Systems Conference on Power Engineering, LESCOPE-04
18. Abyaneh H, Al-Dabbagh M, Karegar H, Sadeghi S, Khan R (2003) A new optimal approach for coordination of overcurrent relays in interconnected power systems. IEEE Trans Power Deliv 18(2):430–435
19. Mohagheghi S, Stoupis J, Wang Z (2009) Communication protocols and networks for power systems-current status and future trends. In: Power Systems Conference and Exposition, PSCE '09. IEEE/PES
20. Clarke GR, Reynders D, Wright E (2004) Practical Modern SCADA Protocols: DNP3, 60870.5 and Related Systems. Elsevier, Oxford
21. Mahat P, Chen Z, Bak-Jensen B (2009) A hybrid islanding detection technique using average rate of voltage change and real power shift. In: Power and Energy Society General Meeting, PES '09. IEEE
22. Fleming B (1998) Negative-sequence impedance directional element. In: 10th Annual Pro Test User Group Meeting, Pasadena, California

Protection System Reliability Assessment Considering Smart Grid Technologies

Ahmed Saleh Alabdulwahab and Roy Billinton

1 Introduction

Smart grid is a new and currently developing platform in the electric power industry. Two of the major characteristics of a smart grid are: utilization of emerging/intelligent devices and improved communications [1]. In recent years, the evolution of modern digital relays has been remarkable. Contemporary digital relays integrate multiple functions (such as metering, protection, automation, control, digital fault recording, and reporting) that efficiently accommodate various power system services. For this reason, they are referred to as intelligent electronic devices (IEDs) [2, 3]. The new IEDs supersede old electromechanical relays and dominate today's power infrastructure. Protection specialists are now given the opportunity to develop innovative communication-based protection schemes using different communication protocols such as the IEC 61850 [4, 5]. With all these advancements, IEDs can be used to augment the integration of protective devices for multiple component items within the substation busbar into one protective relay in which all protective functions in a substation busbar are integrated to form a multifunctional protective device. The incorporation of a robust communication system with the IED creates a powerful tool for power system protection. Upon detecting an event, the IEDs notify those devices that have registered to receive the data. Data can be transmitted within 4 ms from the time an event occurs and are retransmitted multiple times by each IED. The speed of communications between devices depends on the data payload, selected data rate, and the number of network devices. Devices in the peer-to-peer communication network can communicate over radio, fiber optic cables, or twisted pair copper wires. When a fault is detected, the information is transmitted to

A. S. Alabdulwahab (✉)
Electrical and Computer Engineering Department, College of Engineering,
King Abdulaziz University, P.O. Box 80204, Jeddah 21589, Saudi Arabia
e-mail: aabdulwhab@kau.edu.sa

R. Billinton
University of Saskatchewan, 57 Campus Dr.,
Saskatoon, SK, Canada
e-mail: roy.billinton@usask.ca

R. Karki et al. (eds.), *Reliability Modeling and Analysis of Smart Power Systems*,
Reliable and Sustainable Electric Power and Energy Systems Management,
DOI 10.1007/978-81-322-1798-5_5,

all devices in the registry for proper action; providing a high degree of redundancy. Peer-to-peer communication may be implemented in a variety of configurations. It permits sharing of information such as counters, status, voltage or current magnitudes, or angles between the different devices. Each device on the network shares its own data with every other device on the network. The devices in a peer-to-peer communication network take turns sharing their data by broadcasting them on the network [5].

The recent development of nonconventional protection systems and wide spread use of IED supported by advanced communication systems can, however, lead to protection system hidden failures. Hidden failures in protection systems are defined as "a permanent defect that will cause a circuit element(s) to be removed incorrectly and inappropriately as a direct consequence of a switching event [6]." Protection systems consist of many components, such as transducers (current and voltage transformers), relays (R), IED, circuit breakers (CB), and so on, which contribute to the detection and removal of faults [7, 8]. Hidden failures may exist in any of these constituent elements or in the communication system [9, 10].

Hidden failures in the different components of a digitized protection system can affect a wide range of practical protected systems and components. Substations can be configured as single bus, sectionalized bus, main and transfer bus, ring bus, breaker and a half, or double breaker double bus topologies. These are "breaker oriented" configurations. Hidden failures of a protection system can cause a breaker to not open when required. In some of the above noted configurations, a failure of one breaker to open when required may constitute a failure state for the system. In other configurations, the failure of two breakers to open when required creates a failure state. The failure state of a system is dependent on the configuration of the substation, the load conditions, operating policies, and other factors. The probability of a breaker responding or not to a fault depends on the protection system, its construction, and the quality of the components used. The protection system is a completely integrated system of its own which can be analyzed independently of the power system network or component which it is intended to protect. This independent analysis enables sensitivity and comparative studies to be made of alternative technologies and protection schemes which can be employed in smart grids.

Conventionally, the reliability assessment of distribution systems and substations with different configurations such as that presented in the IEEE Std 493-2007 [11] and IEEE Std 1366-2003 [12] is assessed without considering the impact of the hidden failures of protection systems. In some cases the effect of a stuck breaker is incorporated in the analysis [13, 14]. However, this analysis may not be accurate enough for smart grids in modern cities and industrial facilities with extensive deployment of complicated digitized protection systems supported by advanced communication systems. This chapter studies the impact on the reliability of a protection system of the different components of an advanced protection scheme. The outcome of this study can feed into reliability analysis for smart grids or specific substation configurations.

2 Protection System Reliability Assessment

Considerable work has been done to examine different reliability aspects of protection systems. Different techniques have been used such as fault-tree (FT) analysis [15–17], reliability block diagrams (RBD) [17, 18], Monte Carlo simulation (MCS) [17], Markov modeling (MM) [10, 17, 19], event tree analysis (ETA) [20], and so on [21]. An FT is a logic diagram that shows potential events affecting system performance and the relationship between potential events. In a complicated system, the ultimate (top) event can be decomposed into different fault events [22]. In the RBD, blocks can be connected through sets of series and parallel connections in order to accurately present a system. A RBD, however, cannot be used for multifunctional adaptive protection schemes. The state-space based MM is different from the others described above in that it is a dynamic approach [23]. With a MM, more calculation time is required for a large system because a complicated state transition matrix must be solved. A MCS does not involve a specific analytical model but utilizes random variables to simulate the physical process [24]. An ETA is a bottom-up analytical technique. Each of the methods described above has its advantages and disadvantages and it is not the objective of this research work to compare these techniques. An ETA is widely used for engineering systems with mitigating features, and sequences of events which lead to the occurrence of specified consequences, following the occurrence of an initiating event. It is appropriate to apply ETA in cases where the successful operation of a system depends on an approximately chronological, but discrete, operation of its subsystems and components, which is the case with the protection system under study. When a fault occurs, the event tree is deduced as a sequence of events involving success and failure of the different components of the protection system. The possible outcomes events are identified and the associated probabilities are calculated for each event using Eq. (1). These events are then aggregated to generate the possible overall outcomes and the probability for a specific outcome is calculated using Eq. (2).

$$P_{Ej} = \prod_{i=1}^{N} P_{Ci} \tag{1}$$

where:

P_{Ej} Probability of occurrence of event j
P_{Ci} Availability/unavailability of component i
N Number of components contributing to event j

$$P_{O_k} = \sum P_{Ej} \quad for \quad P_{Ej} \in O_k \tag{2}$$

where:

P_{O_k} Probability of outcome O_k

Table 1 Possible outcomes for the protection system considered

Outcome	Breaker(s) open when required	Breaker(s) fails to open when required
O_1	B1 and B2	None
O_2	B1	B2
O_3	B2	B1
O_4	None	B1 and B2

Table 2 Reliability data for the protection system components

Component	Availability	Unavailability
Fault detector	0.999990	0.000010
Relay	0.999900	0.000100
IED	0.999997	0.000003
Trip signal device	0.999970	0.000030
Circuit breaker	0.999700	0.000300
Communication system	0.999970	0.000030

IED intelligent electronic devices

3 Alternative Protection Schemes

A basic reliability assessment for a conventional protection system is described in this section. The effect of different smart grid technologies and protection schemes are then examined and presented under different scenarios. The protection system considered in this study is composed of a fault detector device (FD), a relay (R), two trip signal devices (TS1 and TS2), and two circuit breakers (CB1 and CB2). It is assumed that a reliable DC source is used for the control circuit and the communication system. The effect of the DC source can be incorporated in a similar manner if required. The IED and the communication system are introduced in the advanced protection schemes in this study. The possible outcomes considered for this system in the event tree as a response of an initiating fault event are shown in Table 1.

As noted earlier, a failure of one breaker to open when required may be considered as a system failure state in specific system configurations. In other configurations, the failure of two breakers to open when required constitutes a failure state. Two parameters are used to measure the reliability of different protection schemes. These are the probability of one breaker not opening when required and the probability of two breakers not opening when required. Equation (2) is used to calculate these two indices. The unavailability statistics for the protection system components are given in Table 2 [15, 16, 25].

The following scenarios are considered and the event trees for these scenarios are shown in Fig. 1. Only the breakers not opening when required because of hidden failures of the protection system are noted in the figure.

Scenario 1 It is assumed that no communication system is installed. Thus, TS1 actuates B1 and TS2 actuates B2.

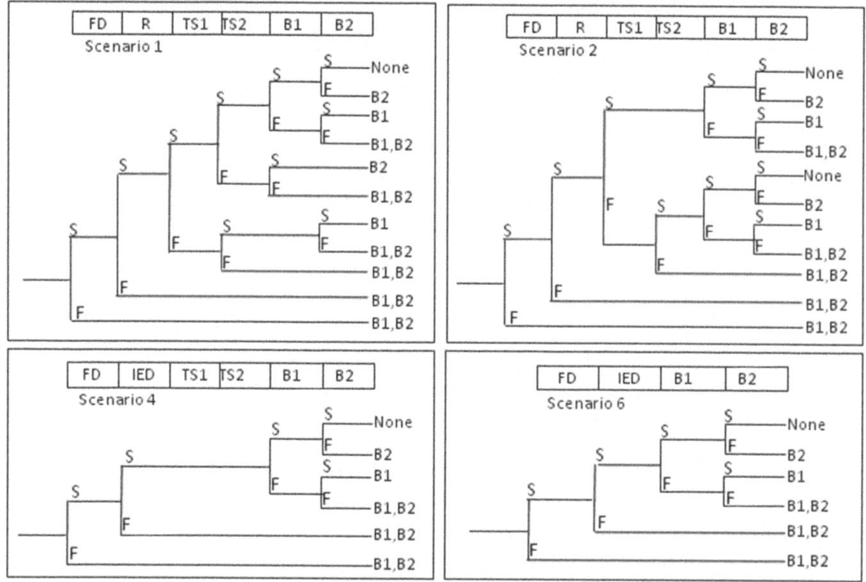

Fig. 1 Event trees for the different scenarios

Scenario 2 It is assumed that a 100% peer-to-peer communication system is installed. Thus, both breakers can be actuated by either one of the TS devices.

Scenario 3 The unavailability of the communication system is considered in this case using the unavailability shown in Table 2. When the communication system fails, the advanced protection system is disabled entirely by switching to the traditional operating mode which leads the system to operate as in Scenario 1. When the peer-to-peer communication system operates successfully, the protection scheme is identical to that in Scenario 2. The probabilities associated under these conditions are calculated using the conditional probability method. Equation (3) is applied using the results obtained in Scenarios 1 and 2.

$$P_{O_k} = \left(P_{O_k} \mid ComS\right) \times A_{com} + \left(P_{O_k} \mid ComF\right) \times U_{com} \tag{3}$$

where:

$(P_{O_k} \mid ComS)$ Probability of outcome k given that the communication system is successful

$(P_{O_k} \mid ComF)$ Probability of outcome k given that the communication system is failed

A_{com} Availability of the communication system

U_{com} Unavailability of the communication system

Table 3 Outcome probabilities for the different scenarios

Breaker not	Scenario						
opening	1	2	3	4	5	6	7
B1	4.3995	4.0997	4.0997	3.1300	3.1300	3.1300	3.4299
B2	4.3995	4.0997	4.0997	3.1300	3.1300	3.1300	3.4299
B1 and B2	1.1011	1.1009	1.1009	0.13090	0.13093	0.13090	0.43090

The numbers shown have been multiplied by 1e4

Scenario 4 The traditional relay is replaced by an IED that is logically programmed to perform as a multifunctional device with a 100 % reliable communication system. In this case, the IED performs the functions of the relay, TS1 and TS2 if any of them fail. The system, therefore, does not fail if both TS1 and TS2 fail with the IED operating successfully.

Scenario 5 The traditional relay is replaced by an IED as in Scenario 4 with the communication system considered to have the unavailability shown in Table 2. If the peer-to-peer communication system fails to operate, the protection system is switched to a contingency scheme in which the IED functions as a relay only and the protection system operates as in Scenario 1. If the peer-to-peer communication system operates successfully, the protection scheme is identical to that in Scenario 4. The probabilities associated under these conditions are calculated using the conditional probability method. Equation (3) is used with the results obtained in Scenarios 1 and 4.

Scenario 6 The traditional relay and the two TS devices are replaced by a multifunctional IED that is logically programmed with a 100 % reliable communication system. In this case, the IED operates as the R, TS1, and TS2.

Scenario 7 The traditional relay and the two TS devices are replaced by an IED that is logically programmed with the communication system and has the unavailability shown in Table 2. If the peer-to-peer communication scheme fails then the IED cannot operate as a TS device and both breakers will fail to trip. Therefore, the probabilities that B1 fails to open, B2 fails to open and both fail to open is 1.0 while the communication system is down. The probabilities of failure are identical to those in Scenario 6 under the condition that the communication system is operating. The probabilities of this scenario are obtained using Eq. (3).

The event trees shown in Fig. 1 and the data given in Table 2 are used to generate the reliability parameters for the different scenarios using Eqs. (1), (2), and (3). The results are shown in Table 3 and Fig. 2.

It can be seen from Fig. 2 that Scenario 1 is the least reliable scheme. In Scenario 2, a reliable communication system is considered, which enables each TS device to communicate with both breakers. This adds redundancy to the system which improves the system reliability. The communication system is assigned a typical unavailability of 3e-5 in Scenario 3. This does not have that much effect on the protection system performance due to the high reliability of the communication system. One more level of redundancy is introduced in the system by replacing the traditional relay by a multifunctional logic programmable IED in Scenario 4. The

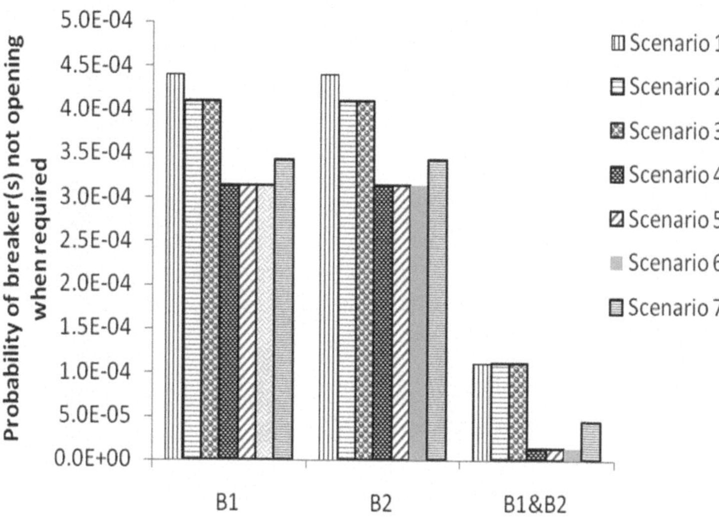

Fig. 2 Probabilities of breaker(s) not opening for the different scenarios

IED is programmed to function as a relay and as the two TS devices. This increase in redundancy is reflected in an increase in the protection system reliability. Considering the unavailability of the communication system in Scenario 5 did not have that much effect for the reasons noted previously. In Scenario 6, both TS devices are removed which reduces the redundancy in the system. Surprisingly, an identical result was achieved with Scenario 4. This is due to the typical high availability of the IED keeping in mind that the communication system is assumed to be 100 % reliable in this scenario. Considering a relatively low unavailability of the communication system in Scenario 7 results in a decrease in system reliability compared to the other improvement scenarios. This is due to the high dependability on the communication system in this case. Scenario 7, however, shows a lower risk compared to the original protection system configuration. Implementing smart grid technologies such as IEDs and robust communication systems in a protection system can have a positive impact on the system reliability. It is important to note, however, that redundancy is advised to mitigate the failure of the IED and the communication system.

The percentage improvements in system reliability (decreased unavailability) compared to Scenario 1 for the different scenarios are shown in Fig. 3.

Figure 3 shows that an improvement of approximately 7 % is achieved with Scenarios 2 and 3 for the one breaker not opening when required case. No discernable improvement is achieved in the case of the two breakers not opening when required for Scenarios 2 and 3, in which the communication system is installed on the traditional protection system. Scenarios 4, 5, and 6 result in a significant reduction in system risk of approximately 30 and 90 % for one breaker and two breakers not opening when required respectively. The utilization of the IED together with the communication system has a major impact on protection system reliability improvement. Scenario 7 results in a major reduction in system risk of

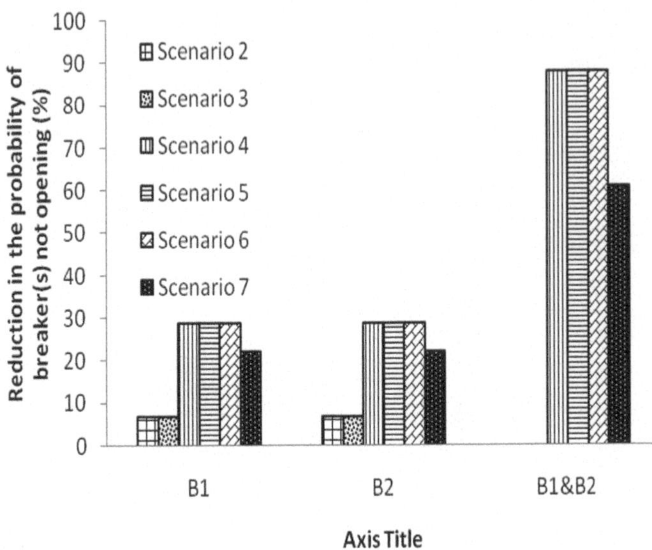

Fig. 3 Reduction in system risk for the different scenarios

approximately 22 and 60 % for one breaker and two breakers not opening when required respectively. This reduction is less than that obtained with Scenarios 4, 5, and 6 due to the reduced redundancy in Scenario 7. A balance between the cost of the redundancy and the required reliability level should be identified.

The previous analysis shows that the advanced smart technologies when integrated in the system result inconsiderable reliability improvement. Designing a new distribution network or the reinforcement of an existing network that is breaker oriented can benefit from the integration of advanced technologies such as IEDs and communication systems. However, the magnitude of reliability improvement with these technologies depends on the system configuration, the protection system construction, and the quality of the components used.

4 Sensitivity Analysis

In the results presented in the previous section, it was assumed that the availabilities of the different protection system components are as given in Table 2. Sensitivity analysis was conducted for the seven described scenarios to assess the protection system reliability with variation in component reliability. The reliability of each component is expressed by its unavailability and the reliability of the protection system is expressed as the probability of B1 not opening when required and the probability of both B1 and B2 not opening when required. In these analyses, the availability of a given component is changed and the other components are assumed to have the initial values shown in Table 2. Figures 4 and 5, respectively, show the

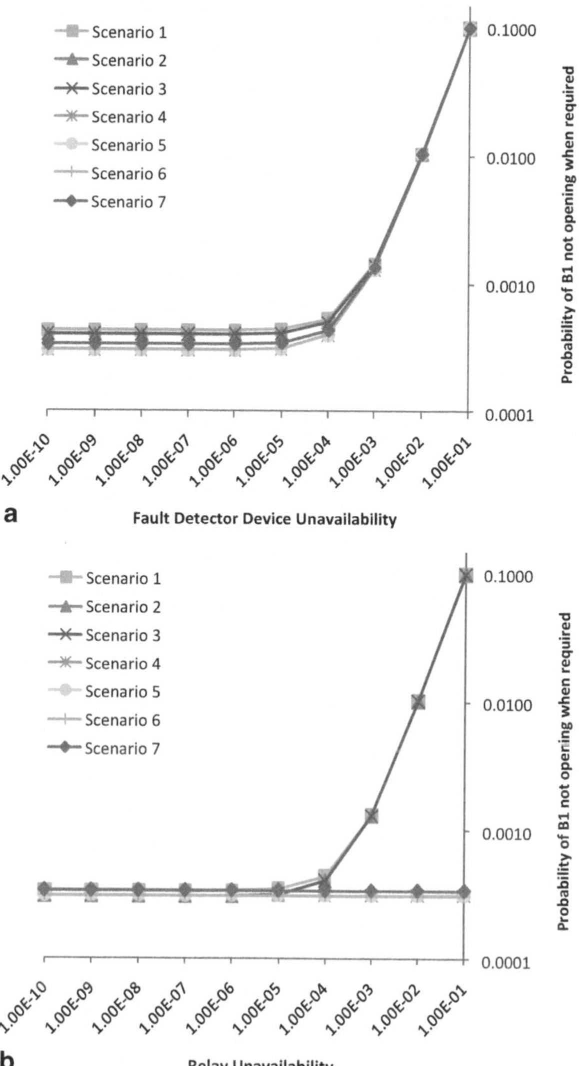

Fig. 4 Effects of different protection system component unavailability on the probability of B1 not opening. **a** Fault detector device unavailability. **b** Relay unavailability. **c** Trip signal devices unavailability. **d** Circuit breakers unavailability. **e** IED unavailability. **f** Communication system unavailability

effects on the probability of B1 not opening when required and the probability of the both B1 and B2 not opening when required with variation in the unavailability of the FD, R, IED, TS device, CB, and the communication system.

FD devices and CB are major components in a protection system and no redundant components are included in the studied scenarios. The unreliability of the protection system for all scenarios increases with a continuous increase in the CB and FD device unavailabilities for all scenarios as shown in Figs. 4a, d and 5a, d. Figures 4b and 5b show that the protection system risk increases with the continuous increase in the relay's unavailability due to the high dependability of the system on this crucial component in Scenarios 1, 2, and 3. Scenarios 4, 5, 6, and 7 are not

Fig. 4 (continued)

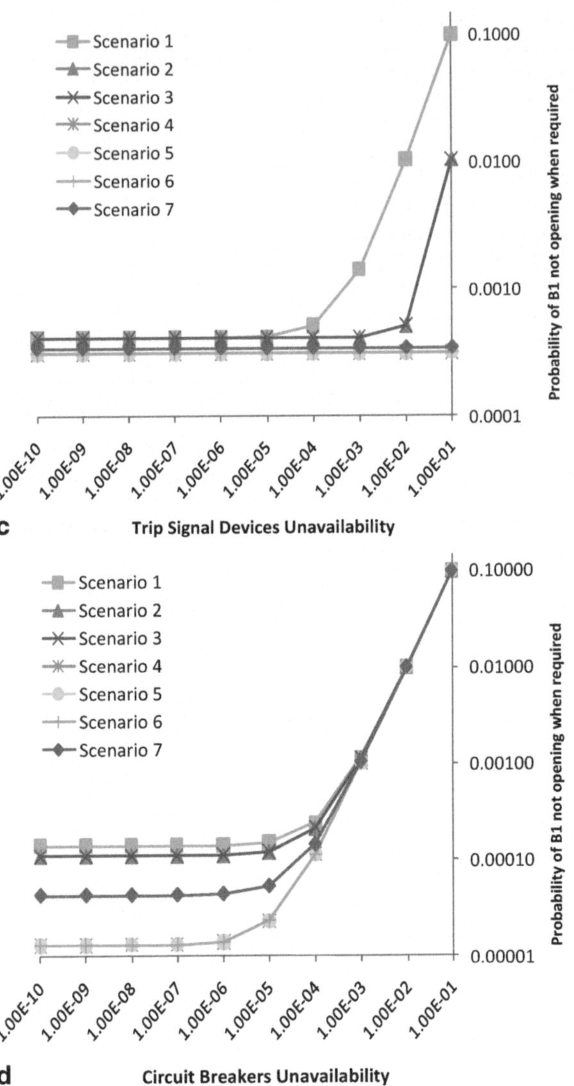

affected by this variation because the relay was replaced by an IED except for Scenario 5 in Fig. 5b which shows an insignificant increase in system risk due to the protection scheme considered. This reflects the improvement in system reliability with the IED installed in a configuration with a highly reliable communication system.

Figure 4c shows that the probability of B1 not opening in Scenario 1 is not affected by the initial increase in TS device unavailability. The probability of B1 not opening in Scenario 1, however, becomes sensitive with further increase in the TS device unavailability to the extent that this probability is almost equal to the unavailability of the TS. This is due to the fact that there is no backup for TS1 if it

Fig. 4 (continued)

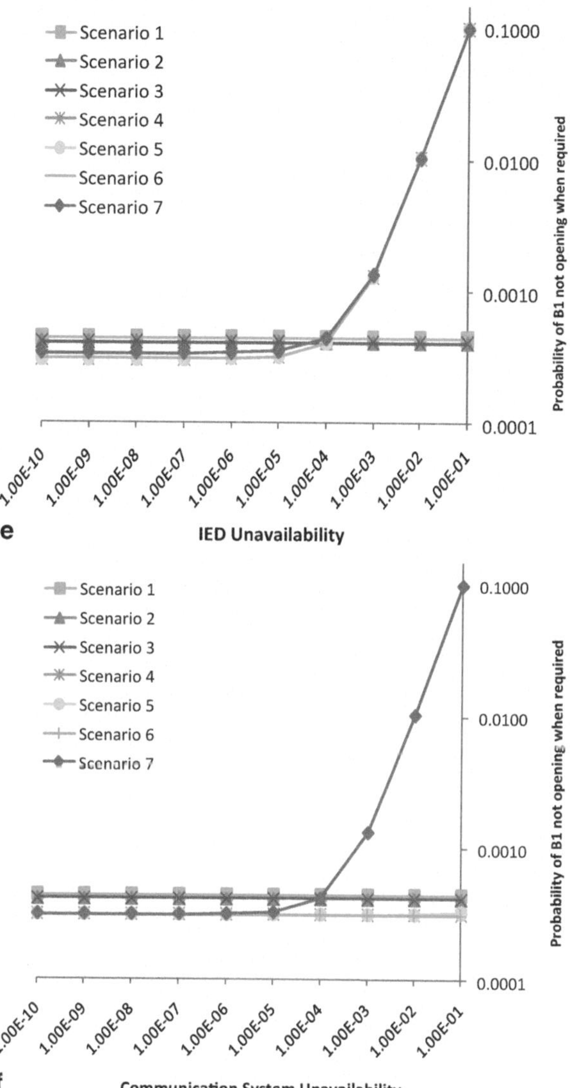

fails. Supporting the protection system with the communication system in Scenarios 2 and 3 improves the redundancy of the protection system. Thus, B1 can be actuated by either one of the TS devices and this is reflected in the reduced risk in Scenarios 2 and 3 compared to Scenario 1. Figure 5c shows a similar reaction to the increase in TS device unavailability. Scenarios 4 and 5 show no reaction with the increase in TS device's unavailiability, which reflects the effectiveness of the redundancy added to the protection system with the IED installed with a high reliable communication system. Scenarios 6 and 7 are not affected in this case because the TS devices are not part of the protection system.

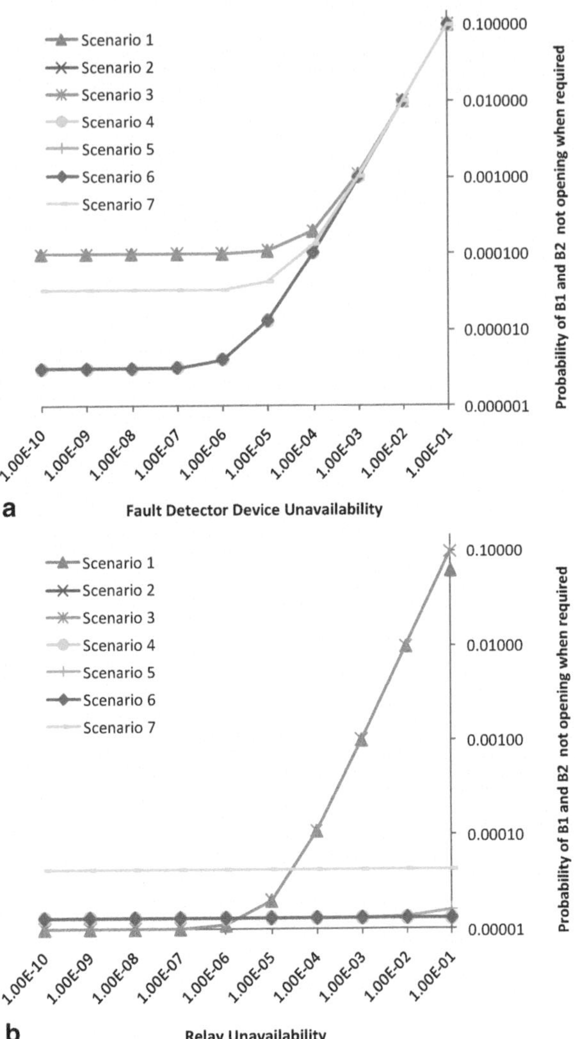

Fig. 5 Effects of different protection system component unavailability on the probability of B1 and B2 not opening. **a** Fault detector device unavailability. **b** Relay unavailability. **c** Trip signal devices unavailability. **d** Circuit breakers unavailability. **e** IED unavailability. **f** Communication system unavailability

Figures 4e and 5e show that the protection system risk increases with the continuous increase in the IED's unavailability due to the high dependability of the system on this crucial component in Scenarios 4, 5, 6, and 7. Scenarios 1, 2, and 3 are not affected by this variation because the relay is used in these cases.

Figures 4f and 5f show that a fully digitized protection system represented by Scenario 7 is very sensitive to the availability of the communication system.

Figure 3 shows that having the IED installed in the protection system reduces the risk of one breaker not opening by approximately 30 % in Scenarios 4, 5, 6 and by 22 % in Scenario 7, and the risk of both breakers not opening by approximately 90 % in Scenarios 4, 5, 6 and by 60 % in Scenario 7. Figures 4e and 5e show that an

Fig. 5 (continued)

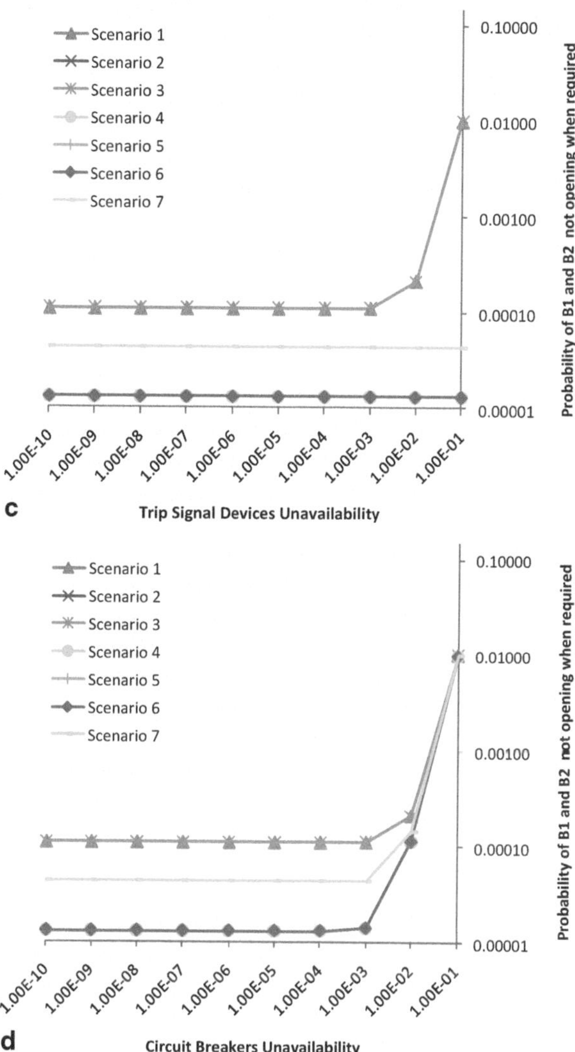

IED unavailability in excess of 1e-4 results in deterioration of the system reliability compared to Scenario 1.

Figure 3 shows that having the communication system support the protection system reduces the risk of one breaker not opening by approximately 7 % in Scenarios 2 and 3 and by 30 % in Scenarios 4, 5, 6, and 7, and the risk of both breakers not opening by only 0.02 % in Scenarios 2, and 3 and by 90 % in Scenarios 4, 5, 6, and 7. This reduction in risk is dominated by the IED in Scenarios 4 to 7. Figures 4f and 5f shows that further increase in the unavailability of the communication system did not increase the probability of a breaker(s) not opening except in Scenario 7 which depends completely on the communication system. A communication system

Fig. 5 (continued)

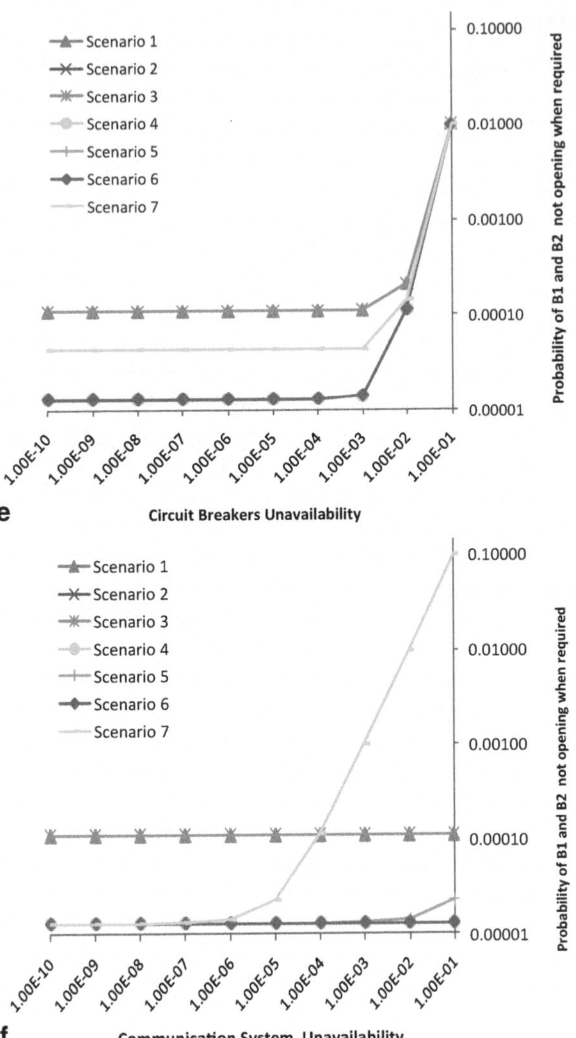

unavailability in excess of 1e-4 results in a deterioration of the system reliability compared to Scenario 1.

The studies show that a digital protection system supported by IEDs and a reliable communication system will result in a high performance protection system provided that a well designed maintenance program is utilized to keep the reliability of the IEDs and the communication system at acceptable levels.

Another sensitivity analysis approach is considered using the so called "measures of importance" [26]. It provides the "influence measure" of a fault event. A larger measure of importance indicates the corresponding fault event is more important compared to other events. A measure of importance is very useful to identify the weak points and components in the system. Once identified, the reliability of

Table 4 RRW using the probability of B1 not opening

	Scenario						
Component	1	2	3	4	5	6	7
FD	1.0232	1.0249	1.0249	1.0329	1.0329	1.0329	1.0300
R	1.2940	1.3225	1.3225	1	1.0000	1	1
TS	1.0731	1.0000	1.0000	1	1.0000	1	1
B	3.1426	3.7269	3.7269	24.0767	24.0699	24.0767	7.9765
IED	1	1	1	1.0097	1.0097	1.0097	1.0088
Comm	1	1	1.0000	1	1.0000	1	1.0958

FD fault detector, *R* relay, *TS* trip signal device, *B* breaker, *IED* intelligent electronic devices, *Comm* communication system

Table 5 RRW using the probability of B1 and B2 not opening

	Scenario						
Component	1	2	3	4	5	6	7
FD	1.0999	1.0999	1.0999	4.2362	4.2332	4.2362	1.3022
R	10.8922	10.9098	10.9098	1	1.0002	1	1
TS	1.0001	1.0000	1.0000	1	1.0000	1	1
B	1.0009	1.0008	1.0008	1.0069	1.0069	1.0069	1.0021
IED	1	1	1	1.2973	1.2972	1.2973	1.0748
Comm	1	1	1.0000	1	1.0002	1	3.2918

FD fault detector, *R* relay, *TS* trip signal device, *B* breaker, *IED* intelligent electronic devices, *Comm* communication system

a component may be improved by using a higher quality component, introducing redundant components, or by improving the maintainability of the component. The measures of importance can be achieved by Birnbaum, criticality, Fussell-Vesely, risk reduction worth (RRW), risk achievement worth (RAW), and so on [27]. In this chapter, RRW method is considered. Equation (4) is used for RRW calculation considering Outcomes 3 and 4 and the results are shown in Tables 4 and 5 respectively.

$$RRW = \frac{P_{O_k}}{\left(P_{O_k} \mid U_i = 0\right)} \tag{4}$$

where:

U_i Unavailability of component *i*

Table 4 shows the RRW considering the probability of B1 not opening when required. The most critical fault event that influences the system reliability is the circuit breaker failure for all scenarios because of the high unavailability of the circuit breaker compared to the other components and failure of the circuit breaker is a major cause of system failure for specific configurations such as that of a radial distribution network.

Table 5 shows the RRW considering the probability of B1 and B2 not opening when required. The circuit breaker is still the most critical component that influences the system reliability for Scenarios 1, 2, and 3. When redundancy is introduced in Scenarios 4, 5, and 6 criticality is shifted to the nonredundant component which is the FD. The communication system in Scenario 7 became the most critical component due to the fact that protection system will not be functional if the communication system fails.

The presented information in Tables 4 and 5 is valuable for system designers when considering alternative technologies. It is also crucial for system operators when designing maintenance plans. The conclusions of the measure of importance analysis using the RRW method is in line with those obtained in the sensitivity analysis.

5 Conclusions

This work investigates the effects of different alternative digital protection schemes supported by an IED and a communication system. The reliability of each scheme is assessed using event tree analysis. Application to a breaker-oriented system is considered. The probability of a breaker(s) not opening due to hidden failure is used to measure the reliability of the different schemes. Sensitivity analysis is conducted to assess the effect of the protection system component reliabilities. The results clearly illustrate the significant impact on the protection system reliability of the installation of a multifunction IED supported by a robust communication system. The studies presented in this chapter show that the advanced smart technologies when integrated in the system results in significant reliability improvement. Designing a new distribution network or the reinforcement of an existing network that is breaker-oriented can benefit from the integration of advanced technologies such as IEDs and communication systems. However, the magnitude of reliability improvement with these technologies depends on the system configuration, the protection system construction, and the quality of the components used. The chapter illustrates that high penetration of advanced technologies in protection systems must also be associated with highly reliable components for enhanced system performance. The outcome of studies such as these can be utilized in the reliability analysis of smart grids or specific substation configurations.

Acknowledgement This work was funded by the Deanship of Scientific Research (DSR), King Abdulaziz University, Jeddah, under grant no. (8-135-D1432). The authors, therefore, acknowledge with thanks to DSRs technical and financial support.

References

1. EPRI (2010) Smart grid leadership report: global smart grid implementation assessment. Final report. October 2010
2. Apostolov AP (2001) Modeling of legacy intelligent electronic devices for UCA based substation integration systems. In: Proceedings of the large engineering systems conference on power engineering, Halifax, Canada, pp 38–43

3. Apostolov A, Muschlitz B (2003) Object modeling of measuring functions in IEC 61850 based IEDs. In: Proceedings of the IEEE transmission and distribution conference and exposition, vol 2, Dallas, USA, pp 471–476
4. Xyngi I, Popov M (2010) IEC61850 overview—where protection meets communication. In: Proceedings of the 10th IET international conference on developments in power system protection, Manchester, UK, 1–5
5. Kojovic LA, Day TR (2003) Advanced distribution system automation. In: Proceedings of the IEEE transmission and distribution conference and exposition, vol 1, Dallas, USA, 348–353
6. De La Ree J, Liu Y, Mili L, Phadke AG, DaSilva L (2005) Catastrophic failures in power systems: causes, analyses, and countermeasures. Proc IEEE 93(5):956–964
7. Horowitz SH, Phadke AG (2008) Power system relaying, 3rd edn. Wiley, UK
8. Zafiropoulos EP, Dialynas EN (2007) Methodology for the optimal component selection of electronic devices under reliability and cost constraints. Qual Reliab Eng Int 23(8):885–897
9. Bozchalui MC, Sanaye-Pasand M, Fotuhi-Firuzabad M (2005) Composite system reliability evaluation incorporating protection system failures. In: Proceedings of the Canadian conference on electrical and computer engineering, Saskatoon, Canada, 486–489
10. Singh C, Patton AD (1980) Protection system reliability modeling: unreadinessprobability and mean duration of undetected faults. IEEE Trans Reliab 29(4):339–340
11. IEEE recommended practice for the design of reliable industrial and commercial power systems. IEEE Std 493–2007 (2007)
12. IEEE guide for electric power distribution reliability indices. IEEE Std 1366–2003 (Revision of IEEE Std 1366–1998 (2004)
13. Billinton R, Allan RN (1996) Reliability evaluation of power systems, 2nd edn. Plenum Press, New York
14. Crispino F, Villacorta CA, Oliveira PRP, Jardini JA, Magrini LC (2004) An experiment using an object-oriented standard-IEC 61850 to integrate IEDs systems in substations. In: Proceedings of the IEEE transmission and distribution conference and exposition, Latin America, pp 22–27
15. Ying-Yi H, Lun-Hui L, Heng-Hsing C (2006) Reliability assessment of protection system for switchyard using fault-tree analysis. In: Proceedings of the international conference on power system technology, Chongqing, China, pp 1–8
16. Schweitzer EO, Fleming B, Lee TJ, Anderson PM (1997) Reliability analysis of transmission protection using fault tree methods. In: Proceedings of the 24th annual western protective relay conference, Washington, USA
17. Tsun-Yu H, Chan-Nan L (2007) Special protection system reliability assessment. In: Proceedings of IEEE industrial & commercial power systems technical conference, Edmonton, Canada, pp 1–7
18. Peichao Z, Portillo L, Kezunovic M (2006) Reliability and component importance analysis of all-digital protection systems. In: Proceedings of the IEEE power systems conference and exposition, Atlanta, USA, 1380–1387
19. Yang W, Wenyuan L, Jiping L (2010) Reliability analysis of wide-area measurement system. IEEE Trans Power Deliver 25(3):1483–1491
20. Kazemi S, Fotuhi-Firuzabad M, Billinton R (2007) Reliability assessment of an automated distribution system. IET Gener Transm Distrib 1(2):223–233
21. IEEE Guide for Selecting and Using Reliability Predictions Based on IEEE 1413. IEEE Std 1413.1–2002 (2003)
22. Tsang AHC (1995) Condition-based maintenance: tools and decision-making. J Qual Maint Eng 1(3):3–17
23. Dhilion BS (2006) Maintainability, maintenance, and reliability for engineers. Taylor & Francis, New York
24. NAVAIR (2005) Guidelines for the naval aviation reliability-centered maintenance process. Patuxent River, Naval Air Systems Command

25. Military Handbook (1991) Reliability prediction of electronic equipment, MIL-HDBK-217F, 2 December 1991
26. Mokhtari A, Frey HC, Danish T (2003) Evaluation of selected sensitivity analysis methods based upon applications to two food safety process risk models. NC State University for US Department of Agriculture, Washington, DC
27. Hsieh CA, Lu CN (2005) Reliability analysis of special protection systems. National Sun-YatSen University for Department of Electrical Engineering, Master Thesis

Smart Charging of Plug-in Electric Vehicles Under Driving Behavior Uncertainty

Marina González Vayá and Göran Andersson

1 Introduction

Plug-in hybrid electric vehicles and electric vehicles (in the following denoted as plug-in electric vehicles or PEVs) have received increasing attention in the field of power systems over the last decade. They are seen both as a challenge and as an opportunity for the power system. On one hand, when charged uncontrolled, they could increase or induce system peaks and lead to the overloading of assets. On the other hand, they could support the system with ancillary services. A considerable part of the literature is concerned with the impact of PEVs on grids and utilities [5, 17, 20], usually coming to the conclusion that some kind of charging control or "smart charging" is needed. Simple schemes such as dual tariffs could be sufficient in early phases but more complex charging schemes, either centralized [13, 23] or decentralized [1, 27] are probably required at larger PEV penetrations [7, 9, 14, 26]. These smart-charging schemes usually rely on the exchange of information between the vehicles and a central entity or aggregator, and are thus linked to the concept of smart grids [6]. Another important part of the literature deals with the fact that vehicles could be used as distributed storage resources to support the system, e.g., providing regulation power or contributing to the integration of renewable energy sources [11, 12, 15]. The concept of an aggregator is crucial to all these schemes. A literature survey on the economic and technical management of this agent can be found in [2] while [19] focuses on regulatory and business aspects. The aggregator would participate in electricity markets to purchase electricity on behalf of the PEVs it manages and either set incentives for vehicles to charge at specific times or manage charging centrally, based on information received from the vehicles. The aggregator could also sell services to the transmission system operator.

M. G. Vayá (✉) · G. Andersson
Power Systems Laboratory, ETH, Zurich, Switzerland
e-mail: gonzalez@eeh.ee.ethz.ch

G. Andersson
e-mail: andersson@eeh.ee.ethz.ch

R. Karki et al. (eds.), *Reliability Modeling and Analysis of Smart Power Systems*, Reliable and Sustainable Electric Power and Energy Systems Management,
DOI 10.1007/978-81-322-1798-5_6, © Springer India 2014

In this chapter, the focus is on smart charging under uncertainty in driving behavior, which is an issue only recently covered in literature, by the authors in [8] and later in [18]. Some papers have already approached the topic of driving behavior stochasticity, however not in the context of charging planning, but instead in that of determining system impacts of uncontrolled charging. In [22], a Monte Carlo approach is chosen to assess the impact of electric vehicles on the distribution network of one of the Azores islands by obtaining average values and confidence intervals for several system indices. In the aforementioned chapter, individual driving behavior is modeled as a time-inhomogeneous Markov chain and charging occurs in an uncontrolled manner. The stochastic process is modeled differently in [16], where variables such as arrival and departure time, and travelled distance are modeled as stochastic variables, and their interdependence is taken into account with copula functions. Queuing theory is the basis for an electric vehicle charging and discharging model in [24] within a probability constrained load flow.

Other papers have focused on smart-charging strategies which rely on perfectly forecasted vehicle behavior. In [13], the optimization problem of an aggregator in a market environment is introduced. Vehicles are clustered with a k-means algorithm and those with similar driving patterns are aggregated. The case where the aggregator is not a price taker is analyzed and the effect of its demand on electricity prices is determined with the help of regression. Possible bottlenecks in the network are not taken into account. A centralized aggregator is also considered in [23], which determines the charging schedules of individual vehicles by minimizing the total cost of electricity and respecting individual vehicle constraints. The impact of charging demand on prices is considered in a simplified way only. However, in this case, distribution grid constraints are taken into account. Wang et al. [25] analyze the effect on locational marginal prices of a dealer, managing battery, swapping stations, and acting as a Stackelberg leader, while the independent system operator acts as the follower. In this case, the pool of batteries is assumed to be large enough so that individual vehicle behavior does not affect the strategy of the dealer.

An interesting approach which takes into account the uncertainty in vehicle arrivals and departures when planning the aggregated charging load of a PEV fleet is described in [18]. Aggregated variables such as the energy content of the aggregated fleet and the upper and lower bounds of this parameter are modeled with a first-order Markov process. The optimization problem is formulated as a Markov decision process and solved with approximate dynamic programming. However, the paper does not describe how the Markov process is parameterized based on individual driving behavior. Moreover, prices are considered exogenous and network constraints are not taken into account.

In this chapter, which is an extended version of [8], we propose an approach that determines aggregated day-ahead nodal charging profiles that

- minimize system generation costs
- satisfy network loading constraints
- satisfy driver end-use constraints, given driving pattern uncertainty.

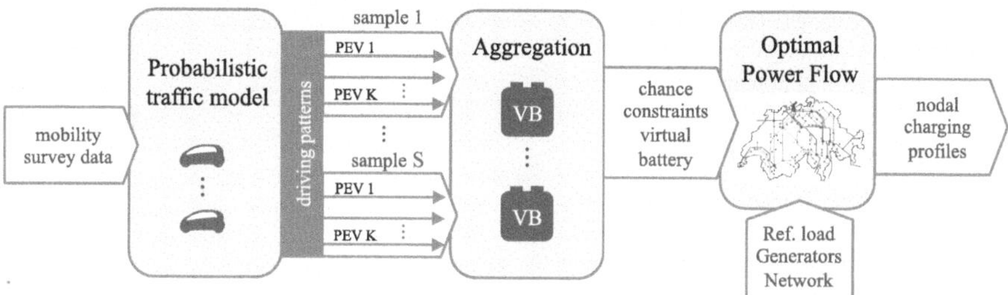

Fig. 1 Schematic representation of the day-ahead charging scheduling approach presented in this chapter

The problem is formulated as a multiperiod DC-OPF with chance constraints. With the OPF framework, the effect of PEV demand, both on prices and on the network, can be taken into account simultaneously. The possibility of discharging energy back to the grid during peak times is not considered here as price arbitrage has been found not to be an economically viable option due to high battery degradation costs [10, 13]. In the OPF, PEVs are modeled in an aggregated way as virtual storage resources at each network node. Stochastic individual driving patterns are created with Monte Carlo simulations based on non-Markov chains, which describe the transition between a driving state and different parking states. With the aggregation of this information, the OPF constraints concerning the virtual batteries can be formulated as chance constraints. Results show that these constraints are violated less frequently with the stochastic formulation than with a deterministic approach.

In the next section, the methodology will be introduced, both concerning the probabilistic transport model and the day-ahead charging scheduling. Then, the simulation setup is described in Sect. 3 and the corresponding results are discussed in Sect. 4. Finally, the conclusion and future work are given in Sects. 5 and 6, respectively.

2 Methodology

A schematic representation of the approach described in this chapter is shown in Fig. 1. As a first step, different samples of individual driving patterns are generated based on mobility survey data, as is detailed in Sect. 2.1. Then, this information is used to model the vehicle fleet as virtual batteries at the different network nodes, whose energy and power bounds are expressed as chance constraints. Within an OPF, the charging profiles of these virtual batteries are determined, as described in Sect. 2.2.

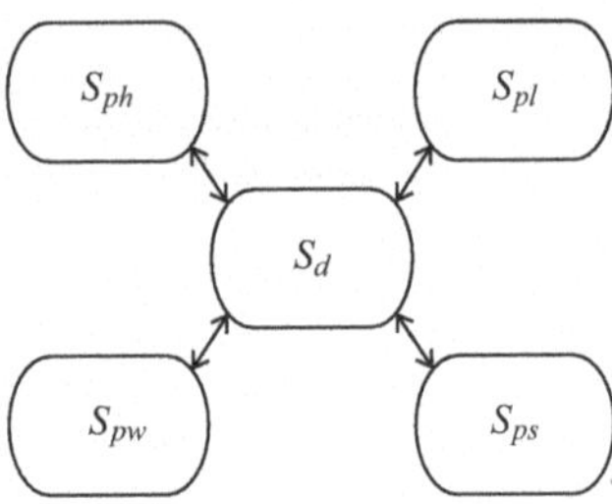

Fig. 2 Possible states and transitions. S_d: driving state, S_{ph}: parked at home, S_{pw}: parked at work, S_{pl}: parked at leisure location, S_{ps}: parked at shopping location

2.1 Probabilistic Transport Model

The goal of the probabilistic transport model is to create possible realizations of mobility patterns out of particular exemplary samples. We define mobility patterns as the set of trips performed with a vehicle as well as the timing, duration, energy consumption, destination, and purpose of these trips.

To model individual driving behavior, five different states are defined: a driving state and four parking states (parked at work, at home, for leisure, or at a shopping location), which are depicted in Fig. 2 together with the possible state transitions. Defining different parked states allows to set some parameters conditional on the state (e.g., only home charging is possible or a higher charging rated power is available at commercial locations than at home). Based on these states and possible state transitions, driving patterns are modeled as a continuous-time non-Markov chain. Unlike in [22], a non-Markov process is chosen because mobility patterns probably do not fulfill the Markov property, i.e., they are not memoryless—the time when a vehicle departs from a particular location is influenced by the time of arrival at that location, and thus by the time of departure from the previous location. In our model, the sojourn time at a particular state is given by two stochastic variables: the time of departure and the trip duration. These variables determine the transition from a parked state to the driving state and the transition from the driving state to a parked state respectively.

It is assumed that on a particular day, a vehicle is used for a certain number of activities performed in a certain order (e.g., home-work-shop-home), which correspond to the parked states. These activity chains are assumed to be perfectly known in advance. The departure time and duration of each trip between activities are modeled as stochastic variables distributed according to a truncated normal distribution whose expected value is known. The truncation is needed to make sure that all activities are performed in the predetermined order. In the next step, the resulting time continuous process is discretized—a vehicle is only considered to be parked at a certain time step (e.g., 15 min or 1 h) if it is there for the whole duration of the time step. Finally, to account for variations in the energy consumed during each trip, this variable is also modeled with a truncated normal distribution.

Available data on driving patterns typically come from national mobility surveys. With these data, distributions of variables such as departure times, etc. for the fleet as a whole can be derived [16], but not distributions for the stochastic behavior of each individual vehicle. Since there is no available data to parameterize the distributions of departure times, etc. for individual drivers, a normal distribution was assumed in this analysis. However, the choice of distribution does not affect the validity of the overall approach presented in this chapter. With real data, it would also be interesting to include correlations, e.g., between the trip duration and the trip energy consumption.

With this framework, a Monte Carlo simulation which leads to different samples of mobility patterns for each vehicle is performed. Based on these, the demand profile of the vehicles can be optimized as shown in the subsequent sections.

2.2 Optimal Charging Scheduling

In this subsection, the following variables will be used:

G_i	Generator $G_i = \{G_1 \ldots G_I\}$
L_j	Load $L_j = \{L_1 \ldots L_J\}$
VB_j	Virtual battery $VB_j = \{VB_1 \ldots VB_J\}$
n	Node $n = \{1 \ldots N\}$
V_k	Vehicle $V_k = \{V_1 \ldots V_K\}$
l_m	Line or transformer $l_m = \{l_1 \ldots l_M\}$
P_{G_i}	Power produced by generator G_i
n_{G_i}	Marginal cost of generator G_i
P_{L_j}	Power consumed by load L_j
$P_{L_j,r}$	Reference load L_j (load w/o vehicles)
$P_{L_j,c}$	PEV load
$P_{V_k,max}$	Maximum power connection of vehicle V_k
C_{V_k}	Battery capacity of vehicle V_k
SOC_{V_k}	Battery state of charge (SOC) of vehicle V_k *min*: minimum SOC *req*: required SOC *trip*: SOC needed for the next trip
E_{VB_j}	Energy content of virtual battery VB_j
$E_{VB_j,d}$	Energy drop from vehicles departing from VB_j
$E_{VB_j,a}$	Energy contribution of vehicles arriving at VB_j
$\overline{\eta}_{VB_j}$	Average charging efficiency of VB_j
$P_{l_m,max}$	Maximum rated power of line or transformer l_m
Ω_{L_j}	Set of V_k associated with total load L_j
Ω_n	Set of G_i and L_j associated with node n
D_{n,l_m}	Power transfer distribution factor associated with line l_m and node n
t	Time step number $t = \{1 \ldots T\}$
Δt	Time step duration

Based on the inputs of the probabilistic transport model and on network, generator and load parameters, the optimal charging profiles can be determined day-ahead. The dispatch of the fleet is performed within a multiperiod optimal power flow framework. For this purpose, PEV characteristics and constraints are aggregated at each node and modeled as a virtual storage resource. The capacity and energy content of these storage resources change over time as vehicles depart/arrive from/to the different nodes. The evolution of the energy content of a virtual battery is given by the charging power, and the energy of the vehicles arriving and departing:

$$E_{VB_j}^{(t)} = E_{VB_j}^{(t-1)} + P_{L_j,c}^{(t)} \cdot \Delta t \cdot \bar{\eta}_{VB_j} - E_{VB_j,d}^{(t)} + E_{VB_j,a}^{(t)} \tag{1}$$

The smart-charging scheme proposed here seeks to minimize total generation costs over the day,

$$\min_{P_{G_i}^{(t)}, P_{L_j}^{(t)}, E_{VB_j}^{(0)}} \sum_{t=1}^{T} \sum_{i} n_{G_i} \cdot P_{G_i}^{(t)} \tag{2}$$

subject to the following constraints:

$$P_{L_j}^{(t)} = P_{L_j,r}^{(t)} + P_{L_j,c}^{(t)} \quad \forall t \tag{3}$$

$$\sum_i P_{G_i}^{(t)} = \sum_j P_{L_j}^{(t)} \quad \forall t \tag{4}$$

$$P_{G_i,min} \leq P_{G_i}^{(t)} \leq P_{G_i,max} \quad \forall t, i \tag{5}$$

$$0 \leq P_{L_j,c}^{(t)} \leq \sum_{V_k \in \Omega_{L_j}^{(t)}} P_{V_k,max}^{(t)} \quad \forall t, j \tag{6}$$

$$\left| \sum_n D_{n,l_m} \cdot (P_{G_i \in \Omega_n}^{(t)} - P_{L_j \in \Omega_n}^{(t)}) \right| \leq P_{l_m,max} \quad \forall t, m \tag{7}$$

$$E_{VB_j,min}^{(t)} \leq E_{VB_j}^{(t)} \leq E_{VB_j,max}^{(t)} \quad \forall j, t \tag{8}$$

$$E_{VB_j}^{(0)} = E_{VB_j}^{(T)} \quad \forall j \tag{9}$$

The first constraint defines the total load as the sum of the non-PEV load, which is considered inelastic in this chapter, and the PEV load. The next group of constraints corresponds to the standard DC-OPF formulation and refers to the power balance in the system (4), the lower and upper bounds of generator (5) and load power (6), and the limits on line and transformer loading (7). The upper bound of the charging power is equal to the total connected power capacity of vehicles at a particular node and time step. In cases where part of the fleet charges in an uncontrolled way, this could also be easily integrated into constraint (6) as a lower bound. The last two constraints correspond to the virtual storage resource—its energy content should not exceed its capacity and not be smaller than a certain lower bound (8) and the initial and final energy content should be equal (9). The lower bound can be determined by the sum of the individual SOC requirements of the vehicles connected.

$$E_{VB_j,min}^{(t)} = \sum_{V_k \in \Omega_{L_j}^{(t)}} SOC_{V_k,req}^{(t)} \cdot C_{V_k} \tag{10}$$

The required SOC of a vehicle is given either by the minimum SOC of the battery or by the SOC needed for the next trip when it is about to depart.

$$SOC_{V_k,req}^{(t)} = \begin{cases} SOC_{V_k,min}, & \text{if } V_k \in \Omega_{L_j}^{(t)} \cap V_k \in \Omega_{L_j}^{(t+1)} \\ SOC_{V_k,trip}^{(t)}, & \text{if } V_k \in \Omega_{L_j}^{(t)} \cap V_k \notin \Omega_{L_j}^{(t+1)} \end{cases} \tag{11}$$

Similarly the upper bound on the energy content of a virtual battery is equal to the sum of batteries capacities connected to the given node,

$$E_{VB_j,max}^{(t)} = \sum_{V_k \in \Omega_{L_j}^{(t)}} C_{V_k} \tag{12}$$

A number of these constraints contain stochastic variables which depend on driving behavior. In (6), the total connected power at a certain node depends on the number of vehicles at that node. The upper and lower bounds in (8) are also stochastic variables—$E_{VB_j,max}$ depends on the number of vehicles connected and their capacity, while $E_{VB_j,min}$ additionally depends on their energy requirements. The evolution of the energy content defined in (1) depends on the departure and arrival of vehicles, so constraints (8) and (9) also depend on this behavior.To formulate the problem in a more compact form, we define the optimization vector as

$$\begin{aligned} x = \Big[& P_{G_1}^{(1)}, \dots P_{G_I}^{(1)}, P_{G_1}^{(2)} \dots P_{G_I}^{(2)}, \dots, P_{G_1}^{(T)} \dots P_{G_I}^{(T)}, \\ & P_{L_1}^{(1)}, \dots P_{L_J}^{(1)}, P_{L_1}^{(2)} \dots P_{L_J}^{(2)}, \dots, P_{L_1}^{(T)} \dots P_{L_J}^{(T)}, \\ & E_{VB_1}^{(0)}, \dots E_{VB_J}^{(0)} \Big]^T \end{aligned}$$

and the cost vector as

$$f = [n_{G_1}, \dots n_{G_n}, n_{G_1}, \dots n_{G_n}, n_{G_1} \dots n_{G_n}, \dots, 0, \dots 0]^T.$$

Then the problem above can be formulated as a stochastic linear program:

$$\min_x f^T \cdot x \tag{13}$$

s.t.

$$A_d \cdot x \leq b, \quad A_s \cdot x \leq \beta \tag{14}$$

$$A_{d,eq} \cdot x = b_{eq}, \quad A_{s,eq} \cdot x = \beta_{eq} \tag{15}$$

where (14) and (15) represent the deterministic and stochastic inequality and equality constraints, with the vectors β and β_{eq} containing stochastic variables. Note that the stochastic variables are only present at the right hand side of the constraints, i.e., the A matrices are deterministic. The link between the two formulations of the problem is detailed as following:

- Deterministic inequality constraint $A_d \cdot x \leq b$ corresponds to (5), (7), and the lower bound of (6)
- Stochastic inequality constraint $A_s \cdot x \leq \beta$ corresponds to (8) and the upper bound of (6), where $E^{(t)}_{VB_j,d}$, $E^{(t)}_{VB_j,a}$, $E^{(t)}_{VB_j,min}$, $E^{(t)}_{VB_j,max}$ and $\sum_{V_k \in \Omega^{(t)}_{L_j}} P_{V_k,max}$ are stochastic inputs
- Deterministic equality constraint $A_{d,eq} \cdot x = b_{eq}$ corresponds to (4)
- Stochastic equality constraint $A_{s,eq} \cdot x = \beta_{eq}$ corresponds to (9), where $E^{(t)}_{VB_j,d}$ and $E^{(t)}_{VB_j,a}$ are stochastic inputs. To obtain the deterministic equivalent of the stochastic problem, the stochastic inequality constraints are defined as separate chance constraints

$$\Pr\left(A(i) \cdot x \leq \beta(i)\right) \geq \alpha(i) \quad \forall i \tag{16}$$

where $A(i)$ and $\beta(i)$ are the rows of A and β respectively, and $\alpha(i)$, the success probabilities imposed on each constraint i. Based on the results of the probabilistic transport model, the cumulative distribution function of each $\beta(i)$ can be approximated and so a value $\beta(i)^*$ can be found so that $\Pr(\beta(i) \leq \beta(i)^*) = 1 - \alpha(i)$ [21]. Then constraint (16) is equivalent to the following constraint:

$$A(i) \cdot x \leq \beta(i)^* \quad \forall i \tag{17}$$

For the deterministic equivalent of the stochastic equality constraints, the median of β_{eq} is used:

$$A_{s,eq} \cdot x = \tilde{\beta}_{eq} \tag{18}$$

By doing this, the expected value of the absolute deviation $| A_{s,eq} \cdot x - \beta_{eq} |$ is minimized.To sum it up, the stochastic linear program has been transformed into

Table 1 PEV parameters

C_{V_k}	$SOC_{V_k,min}$	η_{V_k}	$P_{V_k,max}$
16/24 kWh	0.2	0.9	11 kW

a standard linear program by choosing appropriate values for the right side of the stochastic inequality and equality constraints.

3 Simulation Setup

To validate the described approach, case studies with a model of the Swiss transmission network and a 25 % PEV penetration were performed, which would correspond to a fleet of 1 million PEVs in Switzerland. It was assumed that 50 % of the fleet has a battery capacity of 24 kWh (like the Nissan Leaf) and the other half a capacity of 16 kWh (like the GM Volt and the Mitsubishi iMiEV). The first type was assigned to vehicles traveling larger daily distances. Other important vehicle parameters are also detailed in Table 1. Inputs concerning activity chains, expected departure time, and trip duration and parking location were obtained from the 2005 Swiss mobility survey [3]. Expected consumption was approximated by multiplying trip distances detailed in this survey with an average consumption of 0.2 kWh/km.

Concerning the network, data at the 380/220 kV level were obtained from the Swiss transmission system operator, Swissgrid. The modeled network comprises 191 nodes, 246 lines and 21 transformers. Moreover, data on 134 major generators were collected, and the total load profile corresponding to the load on the Swiss system on a winter weekday, particularly Wednesday, 16 December 2009 [4], was assigned to the individual loads according to the underlying population. Electricity in Switzerland is produced to 95 % by nuclear, run-of-the-river and hydro storage power plants [4]. The marginal costs assumed for these technologies were 17 €/MWh, 0 €/MWh, and 20–100 €/MWh respectively. Since hydro storage power plants are usually the marginal producers in this system, their cost was estimated in terms of opportunity costs according to their average daily production during the season being considered (in this case, wintertime). The time frame chosen for the simulation was 1 day, which was in turn divided into hourly steps.

For the probabilistic transport simulation, a standard deviation of the departure time and trip duration of 5 min and a standard deviation of trip consumption of 1 kWh was chosen. In practice, those values could be chosen based on past observations.

4 Results

First, a thousand Monte Carlo samples were generated with the probabilistic transport simulation, each containing a mobility pattern realization for each vehicle in the fleet. The fraction of vehicles in each of the states over the day, for one of the samples is plotted in Fig. 3. It can be seen that at any given time, more than half of

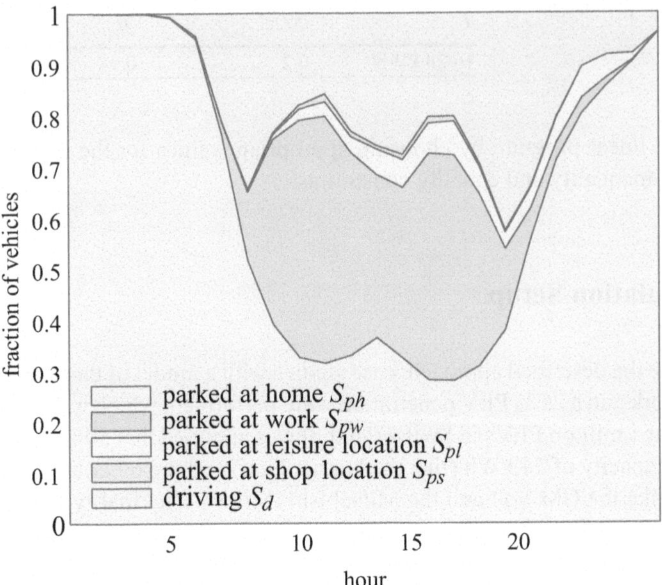

Fig. 3 Fraction of vehicles at each of the states over the day

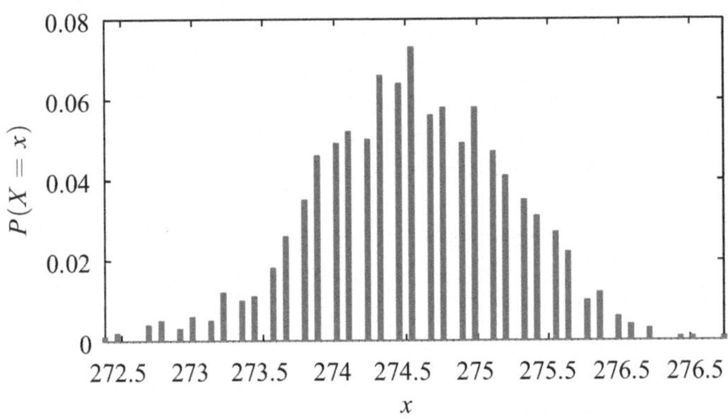

Fig. 4 Probability density function of the sum of connected powers for a given node $L_j = 77$ and time step $t = 12$

the vehicles are parked, which is the reason why PEV load can be considered to be highly flexible.

From the samples of the Monte Carlo simulation, the probability distributions of the stochastic variables $\beta(i)$ associated with the probabilistic inequality constraints (16) were estimated. Figures 4 and 5 show the probability density function and cumulative distribution function of the total connected power at a particular node and time step. It can be seen that the probability density function is discrete and that the

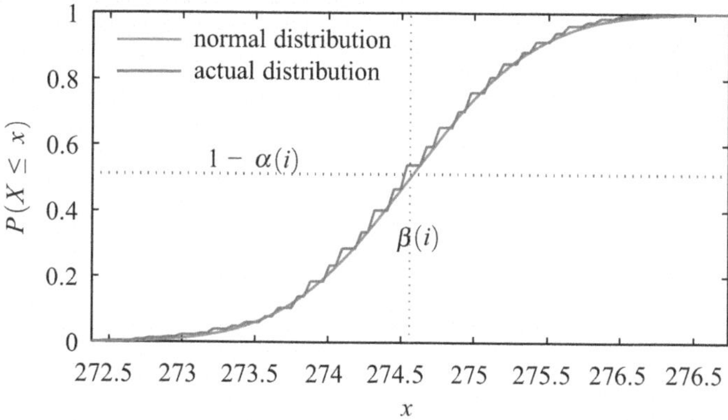

Fig. 5 Cumulative distribution function of the sum of connected powers for a given node $L_j = 77$ and time step $t = 12$

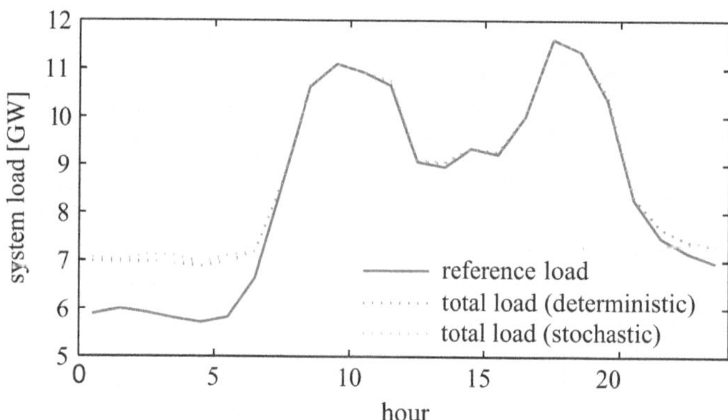

Fig. 6 Load profile without vehicles and resulting load profiles from the multiperiod OPF in the deterministic and stochastic version

cumulative distribution function is close to that of the corresponding normal distribution. Figure 5 also shows how the value $\beta(i)^*$ is chosen out of the cumulative distribution function for a certain $\alpha(i)$.

All $\alpha(i)$ were set to the same value—The probability $\alpha(i)$ was iteratively increased until $\alpha(i) = 0.49$, which is the highest value for which a feasible solution could be found for the given configuration. This implies that there is no feasible solution for a conservative approach where all constraints shall be met in all cases ($\alpha(i) = 1$). Since a robust approach is not possible, the interest of using chance constraints becomes apparent.

The load profile resulting from the stochastic multiperiod OPF is shown in Fig. 6. It can be seen that most charging takes place during valley hours. It is interesting to compare this outcome with the outcome of a deterministic approach, where the

Table 2 Comparison between the deterministic and stochastic approaches

		deterministic	stochastic
$P(a_i^T \cdot x \leq \beta(i))$	min	0	0.49
	avg	0.97	0.99
	max	1	1
$\mathbb{E}(\lvert A_{s,eq} \cdot x - \beta_{eq} \rvert)$	[MWh]	758	43
cost [€]		$9.61 \cdot 10^6$	$9.67 \cdot 10^6$

optimization problem is formulated for the original realization (or equivalently, for the stochastic problem where all standard deviations are set to zero). As can be seen in Fig. 6, the total system load profile is fairly similar in both the stochastic and the deterministic approach. However, the deterministic approach leads to more frequent constraint violations, as detailed in Table 2. On average, the probability of satisfying a particular constraint across different samples is quite high in the stochastic case and only slightly lower in the deterministic case (0.99 and 0.97 respectively). However, this probability is zero for certain constraints in the deterministic approach, while it never falls below the value of $\alpha(i) = 0.49$ in the stochastic approach by design. Moreover, the average absolute deviation of the stochastic equality constraint is also reduced, thanks to the stochastic approach. It can also be seen that the stochastic approach leads to a small increase in costs (0.5 % increase), because it is more constraining.

Finally, Fig. 7 which depicts asset loading, illustrates the importance of including grid constraints in the dispatch of PEVs. It can be seen that, with the constraints, overloading is successfully avoided as expected, that is, loading never exceeds 100 %. However, this is not the case if the constraints are not included, as shown by the dark red peaks appearing in the lower graphic even at valley hours, which correspond to per unit loadings higher than one.

5 Conclusion

Uncontrolled charging could lead to the overloading of system assets, especially at large PEV penetrations. One way to avoid this is the centralized management of charging by an aggregator. Some studies have focused on centralized smart-charging schemes but they usually rely on perfect knowledge of driving behavior and either assume exogenous prices or do not take grid constraints into account. In this chapter, we propose an approach that seeks to overcome these shortcomings. Dispatching vehicles with a multiperiod OPF allows for endogenous prices while implicitly taking into account grid constraints. In this multiperiod OPF, stochastic parameters stemming from driving behavior uncertainty, have been identified and chance constraints have been used to account for this uncertainty. Through a probabilistic transport model, the necessary cumulative distribution functions could

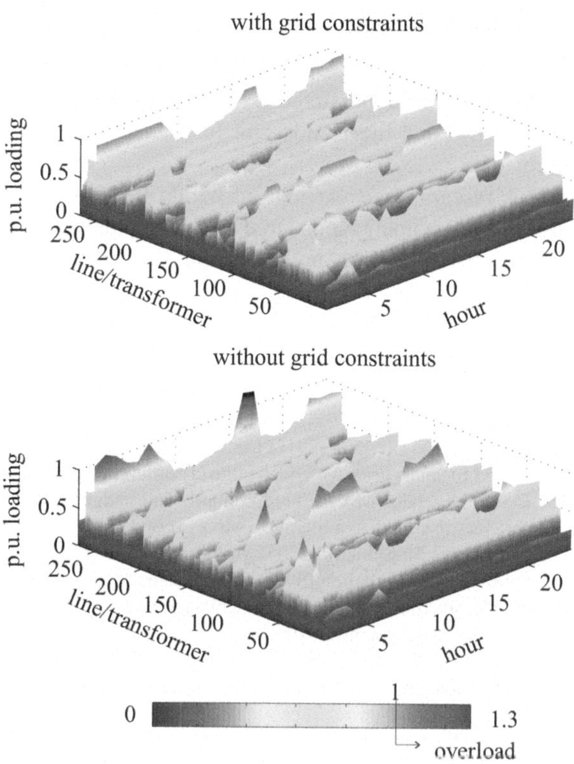

Fig. 7 Line and transformer loading depending on whether the grid constraints (7) are included or not

be estimated to transform the stochastic problem into a deterministic equivalent. Results indicate that the proposed approach reduces the chances of violating the constraints at a small additional cost. It has also been shown that it is necessary to take into account grid constraints when determining charging profiles, even at transmission system level.

6 Future Work

Future work involves moving from the day-ahead scheduling presented here to the dispatch of individual vehicles during operation. For this purpose, Model Predictive Control techniques can be used.

Furthermore, the availability of the fleet for the provision of ancillary services, such as frequency regulation, can be analyzed by extending the proposed framework into a co-optimization of both the goals of cost minimization and reserving enough flexibility to be able to provide a given service. An additional source of uncertainty is introduced here since the energy and power profiles of the ancillary service are typically not known in advance.

7 Acknowledgments

The authors would like to thank Swissgrid for providing the data on the Swiss transmission grid and the Bundesamt für Statistik for providing the detailed mobility survey data.

This research was carried out within the project Technology-centered Electric Mobility Assessment, sponsored by the Competence Center for Energy & Mobility, SwissElectric Research, and the Erdöl-Vereinigung.

References

1. Ahn C, Li CT, Peng H (2011) Decentralized charging algorithm for electrified vehicles connected to smart grid. American control conference (ACC), 2011, pp 3924–3929
2. Bessa RJ, Matos MA (2011) Economic and technical management of an aggregation agent for electric vehicles: a literature survey. Eur Trans Electr Power 22(3):334–350
3. Bundesamt für Statistik (2005) Mobilität in der Schweiz—Ergebnisse des Mikrozensus 2005 zum, Verkehrsverhalten (Swiss transportation statistics, in German), 2005
4. Bundesamt für Energie (2010) Schweizerische Elektrizitätsstatistik 2009 (Swiss electricity statistics, in German), 2010
5. Denholm P, Short W (2006) An evaluation of utility system impacts and benefits of optimally dispatched plug-in hybrid electric vehicles. Technical report, National Renewable Energy Laboratory
6. Galus MD, González Vayá M, Krause T, Andersson G (2012) The role of electric vehicles in smart grids. Wiley Interdisciplinary Reviews, Energy and Environment
7. González Vayá M, Andersson G (2012) Centralized and decentralized approaches to smart charging of plug-in vehicles. IEEE PES general meeting, San Diego, 2012
8. González Vayá M, Andersson G (2012) Smart charging of plug-in vehicles under driving behaviour uncertainty. 12th international conference on probabilistic methods applied to power systems (PMAPS), Istanbul, 2012
9. González Vayá M, Galus MD, Waraich RA, Andersson G (2012) On the interdependence of intelligent charging approaches for plug-in electric vehicles in transmission and distribution networks. IEEE PES innovative smart grid technologies Europe (ISGT Europe), Berlin, 2012
10. González Vayá M, Krause T, Waraich RA, Andersson G (2011) Locational marginal pricing based impact assessment of plug-in hybrid electric vehicles on transmission networks. CIGRE International Symposium, 2011
11. Kempton W, Tomic J (2005) Vehicle-to-grid power fundamentals: calculating capacity and net revenue. J Power Sources 144(1):268–279
12. Kempton W, Tomic J (2005) Vehicle-to-grid power implementation: from stabilizing the grid to supporting large-scale renewable energy. J Power Sources 144(1):280–294
13. Kristoffersen TK, Capion K, Meibom P (2011) Optimal charging of electric drive vehicles in a market environment. Appl Energy 88(5):1940–1948
14. Lopes J, Soares F, Almeida P (2009) Identifying management procedures to deal with connection of electric vehicles in the grid. IEEE Power Tech Conference, 2009
15. Lopes JAP, Almeida P, Soares FJ (2009) Using vehicle-to-grid to maximize the integration of intermittent renewable energy resources in islanded electric grids. International conference on clean electrical power, 2009, pp 290–295
16. Pashajavid E, Golkar MA (2012) Multivariate stochastic modeling of plug-in electric vehicles demand profile within domestic grid. 12th international conference on probabilistic methods applied to power systems (PMAPS), Istanbul, 2012

17. Roe C, Meisel J, Meliopoulos AP, Evangelos F, Overbye T (2009) Power system level impacts of PHEVs. 42nd Hawaii international conference on system sciences (HICSS), 2009
18. Ruelens F, Vandael S, Leterme W, Claessens BJ, Hommelberg M, Holvoet T, Belmans R (2012) Demand side management of electric vehicles with uncertainty on arrival and departure times. IEEE PES innovative smart grid technologies Europe (ISGT Europe), Berlin, 2012
19. San Román TG, Momber I, Abbad MR, Sánchez Miralles A (2011) Regulatory framework and business models for charging plug-in electric vehicles: infrastructure, agents, and commercial relationships. Energy Policy 39(10):6360–6375
20. Schneider K, Gerkensmeyer C, Kintner-Meyer M, Fletcher R (2008) Impact assessment of plug-in hybrid vehicles on Pacific Northwest distribution systems. IEEE power and energy society general meeting—conversion and delivery of electrical energy in the 21st Century, 2008
21. Sen S, Higle JL (1999) An introductory tutorial on stochastic linear programming models. Interfaces 29:33–61
22. Soares FJ, Lopes JAP, Almeida PMR (2010) A Monte Carlo method to evaluate electric vehicles impacts in distribution networks. IEEE conference on innovative technologies for an efficient and reliable electricity supply (CITRES), 2010, pp 365–372
23. Sundström O, Binding C (2012) Flexible charging optimization for electric vehicles considering distribution grid constraints. IEEE Trans Smart Grid 3(1):26–37
24. Vlachogiannis JG (2009) Probabilistic constrained load flow considering integration of wind power generation and electric vehicles. IEEE Trans Power Syst 24(4):1808–1817
25. Wang L, Lin A, Chen Y (2010) Potential impact of recharging plug-in hybrid electric vehicles on locational marginal prices. Nav. Res. Logist. (NRL) 57(8):686–700
26. Waraich RA, Galus MD, Dobler C, Balmer M, Andersson G, Axhausen K (2009) Plug-in hybrid electric vehicles and smart grid: investigations based on a micro-simulation. 12th international conference of the international association for travel behaviour research, 2009
27. Zhongjing M, Callaway D, Hiskens I (2010) Decentralized charging control for large populations of plug-in electric vehicles. 49th IEEE conference on decision and control (CDC), 2010, pp 206–212

Multivariate Stochastic Modeling of Plug-in Electric Vehicles Demand Profile Within Domestic Grid

Ehsan Pashajavid and Masoud Aliakbar Golkar

1 Introduction

Recently, an increasing attention has been paid to plug-in electric vehicles (PEVs) as one of the promising and effective means to cope with the energy crisis and to alleviate the environmental and social concerns [1–2]. The main reasons and motivations for developing PEVs can be briefly summarized as follows:

- To increase the energy security by both reducing a widespread dependence on oil and shifting demand toward non-fossil fuel sources
- To reduce the constantly increasing expenditure on importing and consuming fossil fuels for transportation
- To reduce harmful greenhouse gas emissions and air pollution that leads to climate change
- To accommodate renewable generation to a greater extent
- To represent a responsive and gridable load capable of alleviating peak power demand and provide ancillary services

Growing economic and environmental policy incentives as well as pressure from energy drivers along with technological improvements in battery industry encourages automotive manufacturers to develop various types of PEVs. Taking into account this extensive amount of regulatory and technical efforts focused on the market penetration of the PEVs, it can be anticipated that a numerous number of PEVs will be undoubtedly on the road reshaping the traditional view of power and transportation systems [3–4].

The possibility of bidirectional power transfer with the grid can be regarded as one of the salient characteristics of PEVs. In fact, PEVs can absorb power from the grid when the state-of-charge (SOC) of the battery is low. On the other hand, it is

E. Pashajavid (✉) · M. A. Golkar
K. N. Toosi University of Technology, Tehran, Iran
e-mail: Pashajavid@ee.kntu.ac.ir

M. A. Golkar
e-mail: Golkar@eetd.kntu.ac.ir

R. Karki et al. (eds.), *Reliability Modeling and Analysis of Smart Power Systems*,
Reliable and Sustainable Electric Power and Energy Systems Management,
DOI 10.1007/978-81-322-1798-5_7, © Springer India 2014

possible for these vehicles to inject power into the grid (V2G), especially, during the peak load hours [5]. However, V2G requires a longer time to be practical. PEVs can be generally divided into two categories, namely battery electric vehicles (BEV) which are pure electric vehicles and plug-in hybrid electric vehicles (PHEV) that include both electric motors and internal combustion engines (ICEs) [6].

From grid perspective, modeling the power demand of the PEVs may be considered as one of the critical challenges. This issue can be addressed as deriving of the hourly load consumption of PEVs as well as quantifying the load delivered through a distribution transformer which can be useful for various distribution system applications such as network planning, load management, and probabilistic load flow as well as sitting and sizing issues [7].

Home arrival time, daily travelled distance, home departure time, driving habits, and road traffic conditions may be considered as the most effective factors on the load demand of PEVs [8]. Majority of the previous PEV modeling studies have used deterministic methods to derive the demand profile of the PEVs. However, the above-mentioned factors, due to the related uncertainties, are not deterministic. As a consequence, applying the deterministic methods to model these factors cannot be acceptable and it is essential to employ stochastic and probabilistic modeling approaches [9]. For instance, [10] simply assumes that the PEVs demand profile as well as the number of the connected PEVs to grid follow the univariate normal probability distribution functions (PDF), but there is no reason or data on which this assumption is based. As another example, [11] provides a probabilistic modeling method to estimate the load demand of the PEVs, but the applied method assumes the related factors are independent and there is no correlation among them.

A set of copula-based models are provided in order to extract the hourly aggregated load demand of PEVs delivered through a distribution transformer. Although choosing a distribution for each individual variable may be straightforward, deciding what dependencies should exist among them may not be. Copulas are functions that characterize dependencies among variables and present an approach to create distributions that model correlated multivariate data [12]. Applying a copula, a multivariate distribution can be constructed by specifying marginal univariate distributions and afterwards choosing a copula to provide a dependence structure among variables. Home arrival time, daily travelled distance, and home departure time of a set of commuter internal combustion vehicles in Tehran are utilized as the input datasets. In order to generate the required synthetic data, the aforesaid random variables (RV) are, at first, modeled as marginal distributions. The univariate marginal distribution can be chosen according to its features. The copula function does not constrain choice of the marginal distribution. Unlike most of the recent researches that used the normal PDF for the sake of simplicity [10–11], appropriate non-Gaussian PDFs are suggested to fit to the RVs. In addition, the effectiveness of the modeled PDFs in comparison with the normal PDF is illustrated. Then, the dependence structure is modeled using a student's *t* copula distribution. Afterwards, the Monte Carlo-based stochastic modeling algorithm with two scenarios to extract the initial state-of-charge (SOC_{init}) of PEV batteries is thoroughly explained. By applying the approach, the hourly load distribution functions of the PEVs is derived. Eventually, the distributions are utilized to extract the demand profile of the fleet.

The remainder of this chapter is organized as follows. The notion of copula and the algorithm to construct a copula PDF is explained in Sect. 2. Section 3 elaborates the proposed stochastic algorithm. Afterwards, the simulation results are discussed in Sect. 4. Finally, the subject is concluded in Sect. 5.

2 Copula

This chapter introduces the notion of a copula function and its probabilistic interpretation, which allows us to consider it a "dependence function." Copulas are functions that characterize dependencies among variables and present an approach to create distributions that model correlated multivariate data [12–13]. Applying a copula, a multivariate distribution can be constructed by specifying marginal univariate distributions, and afterwards, combining the univariate distributions to provide dependence structure. In fact, copula functions C:[0 1]^p–[0 1] are used to relate univariate marginal distributions $F_1(x_1)$, $F_2(x_2)$, …, $F_p(x_p)$ to their joint distribution function $H(x_1, x_2, \ldots, x_p)$ as:

$$C\left(F_1(x_1), F_2(x_2), \ldots, F_p(x_p), \boldsymbol{\rho}\right) = H\left(x_1, x_2, \ldots, x_p\right) \tag{1}$$

where $F_k(x_k) = u_k$, $k = 1, \ldots, p$, and H are the cumulative distribution functions (CDF). It should be stressed that the copula function does not constrain choice of the marginal distributions.

Several assessments have been carried out concerning the construction of different families of copulas and their properties [14–15]. Consequently, various kinds and families of copula such as the Elliptical copulas and the Archimedian copulas have been presented and investigated. The student's t copula (t copula) is a popular and well-known member of elliptical copulas that is employed in this chapter ([16], also available at: www.mathworks.com). The student's t copula is considered to be an appealing alternative to the normal copula. Generally, the multivariate student's t distribution is more suitable for real world data than normal distribution, particularly because its tails are heavier and therefore realistic. Equation (2) shows the univariate student's t PDF, where $\Gamma(.)$ is the Gamma function and υ is the degrees of freedom. The multivariate student's t PDF, h_t, is parameterized with $\boldsymbol{\rho}$, the linear correlation matrix, $\boldsymbol{\mu} = [\mu_1, \mu_2, \ldots, \mu_p]^T$, mean vector and υ, the degrees of freedom.

$$f_t(x) = \frac{\Gamma\left(\frac{\upsilon+1}{2}\right)}{\Gamma\left(\frac{\upsilon}{2}\right)\left(1+\frac{x^2}{\upsilon}\right)^{\frac{\upsilon+1}{2}}\sqrt{\pi\upsilon}} \tag{2}$$

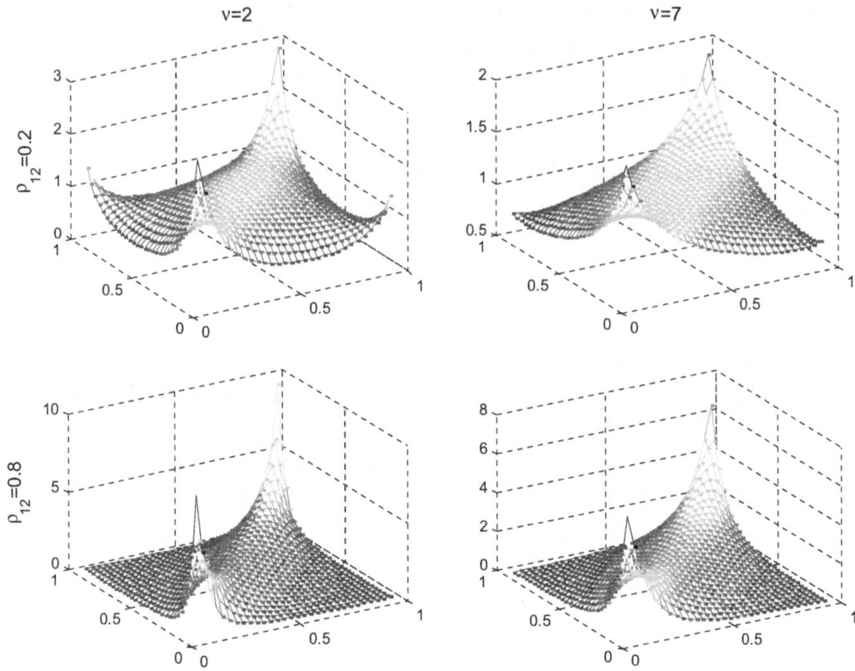

Fig. 1 Plots of the bivariate student's *t* copula PDFs

Let $\boldsymbol{\rho}$ be a symmetric, positive definite matrix with unity diagonal members and H_t the standardized ($\boldsymbol{\mu}=0$) student's *t* joint CDF:

$$h_t(\mathbf{x}) = \frac{\Gamma\left(\dfrac{\upsilon+p}{2}\right)|\boldsymbol{\rho}|^{-\frac{1}{2}}(\upsilon\pi)^{\frac{p}{2}}}{\Gamma\left(\dfrac{\upsilon}{2}\right)\left(1+\dfrac{1}{\upsilon}(\mathbf{x}-\boldsymbol{\mu})^T\boldsymbol{\rho}^{-1}(\mathbf{x}-\boldsymbol{\mu})\right)^{\frac{\upsilon+p}{2}}} \tag{3}$$

$$H_t(x_1,x_2,\ldots,x_p)=\int_{-\infty}^{x_1}\int_{-\infty}^{x_2}\ldots\int_{-\infty}^{x_p}h_t dx_1 dx_2\ldots dx_p \tag{4}$$

Then, for any u$=(u_1, u_2,\ldots, u_p)\in[0\ 1]^p$ the student's *t* copula is defined as follows:

$$\begin{aligned} C_t\left(u_1,\ldots,u_p\right) &= H_t\left(F_t^{-1}\left(u_1\right),F_t^{-1}\left(u_2\right),\ldots,F_t^{-1}\left(u_p\right)\right) \\ &= \int_{-\infty}^{F_t^{-1}(u_1)}\int_{-\infty}^{F_t^{-1}(u_2)}\ldots\int_{-\infty}^{F_t^{-1}(u_p)}h_t dx_1 dx_2\ldots dx_p \end{aligned} \tag{5}$$

where F_t^{-1} is the inverse of the univariate CDF of student's *t* with υ degrees of freedom. Figure 1 shows the PDFs for the bivariate student's *t* copula with $\rho_{12}=0.2$

and $\rho_{12}=0.8$ with $\upsilon=2$ and $\upsilon=7$. It is worthy to notice that these PDFs are merely concerned with the correlation attributes of the distributions.

The random data generation according to the original datasets is required for the Monte Carlo simulation. In order to generate the synthetic data utilizing the student's *t* copula distribution, the following algorithm may be applied:

- Fit marginal distributions to the input dataset x_k, $k=1,\ldots,p$.
- Use appropriate CDF functions to transform $[x_1, \ldots, x_p]$ to corresponding u_k:[0, 1], $k=1,\ldots,p$.
- Use inverse student's *t* CDF function to transform $[u_1, \ldots, u_p]$ to *t* distribution, $[x_1^t,\ldots,x_p^t]$.
- Estimate the linear correlation $\boldsymbol{\rho}$ among the datasets $[x_1^t,\ldots,x_p^t]$ and then, generate determined number of random samples $[\tilde{x}_1^t,\ldots,\tilde{x}_p^t]$ according to the fitted copula as (5).
- Use student's *t* CDF function to transform $\left[\tilde{x}_1^t,\ldots,\tilde{x}_p^t\right]$to the new data $\tilde{u}_k$:[0, 1], $k=1,\ldots,p$.

Use inverse CDF functions appropriate to the related marginal distribution in order to transform $[\tilde{u}_1,\ldots,\tilde{u}_p]$ to the output datasets $[\tilde{x}_1,\ldots,\tilde{x}_p]$.

3 The Modeling Methodology

3.1 *The Proposed Algorithm*

It is presumed that only the home charging is available and the PEVs are plugging in the grid to be charged as soon as they arrive at home. The overall procedure of the proposed approach is visible in Fig. 2. The datasets include home arrival time ($\mathbf{a}_t$), daily travelled distance ($\mathbf{tr}_d$), and home departure time ($\mathbf{d}_t$) of the vehicles during weekdays. At first, the random samples of $\mathbf{a}_t$ (a_n), $\mathbf{tr}_d$ (tr_n), and $\mathbf{d}_t$ (d_n) should be generated by using the extracted *t* copula function. The available charging time (t_{avi}) for each of the PEVs is calculated by subtracting the departure time of the next day (d_{n+1}) from the arrival time of today (a_n). The battery capacity (Cap_{bat}), SOC_{init}, power rating (P_{rat}), and efficiency (C_{chr}) of the battery chargers determine the necessary time (t_{full}) to fully charge the battery. In case t_{full} is less than t_{avi}, the full charging of the battery is accomplished. Otherwise, it is impossible to fully charge the battery. The bigger the battery capacity and the lesser the power rating of the charger, the longer is the time required to fully charge the battery. Accordingly, the hourly power consumption of the PEVs (DPIP_n) is fulfilled in order to estimate the demand profile of the fleet (DPF_m). It is worthy to notice that the PEVs are assumed to be charged through a distribution transformer. Thus, profile of the power delivered to the PEVs through the transformer is attained.

Given the IN as the iteration number of the Monte Carlo simulation, the explained procedure is carried out for IN times in order to derive the distribution of the aggregated power consumption of the fleet within each hour (DAPF_h). Next,

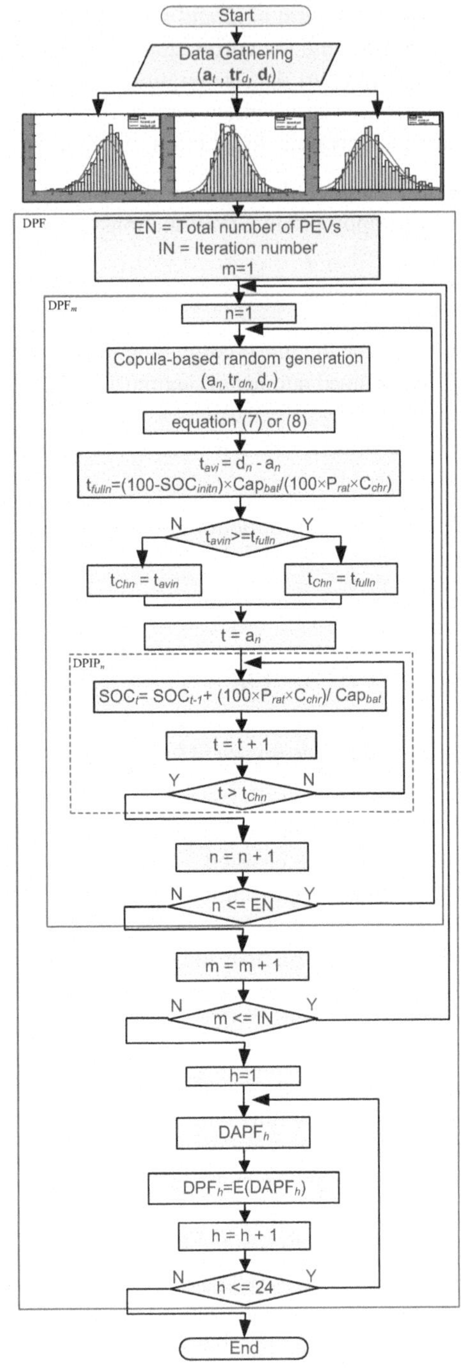

Fig. 2 Flowchart of the proposed approach

the expected values of the hourly load demand of the PEVs can be calculated with respect to the $DAPF_h$. Eventually, demand profile of the fleet (DPF) is estimated by employing the obtained hourly expected values.

3.2 Battery Initial SOC

Extraction of the battery SOC of the PEVs at the charging start time (SOC_{init}) is done with respect to the following scenario cases:

Case 1 As the worst case scenario, the battery initial SOC of the PEVs (SOC_{init}) is supposed to be a constant. This constant is determined according to the depth-of-discharge (DOD) of the PHEVs as follows:

$$SOC_{initn} = 100 - DOD \tag{6}$$

However, the mentioned assumption is not feasible in reality. Therefore, it is suggested to alter (6) by using randomly generated positive numbers by employing an exponential PDF as below:

$$\begin{cases} SOC_{initn} = 100 - DOD + Rand\left(f_{\exp}(x)\right) \\ f_{\exp}(x) = \dfrac{1}{\mu_{\exp}} e^{-\frac{x}{\mu_{\exp}}} \end{cases} \tag{7}$$

Case 2 The SOC_{init} is determined based on the daily travelled distances. Regarding the fact that t_{avi} in home charging is usually bigger than t_{full}, it is rational to assume that the battery SOCs are 100 % at the departure time. Hence, the SOC_{init} of a PEV can be derived as:

$$SOC_{initn} = 100 - \frac{tr_{dn}}{C_{eff} \times Cap_{bat}} \times 100 \tag{8}$$

where C_{eff} is the efficiency coefficient of the PEVs during driving, dependent on the driving patterns and traffic conditions as well as driver efficiency of the electric motors.

4 Simulations and Discussion

Taking into account the assumption that the fleet is delivered power through a distribution transformer, it seems rational to perform the simulation for 20 PEVs. The above-explained procedure is repeated 10,000 times to deal with the related uncertainties. The simulation parameters can be found in Table 1 where C_{chr} stands for

Table 1 The simulation parameters

IN	10,000
EN	20
DOD	70%
P_{rat}	3 kW
Cap_{bat}	20 kWh
C_{eff}	2 km/kWh
C_{chr}	0.9
μ_{exp}	8

efficiency coefficient of a battery charger. At the low voltage domestic level in Iran, the power is delivered through 220 V single-phase connections with a current rating of 16 A. Hence, the power rating of a battery charger through a domestic connection (P_{rat}) is assumed to be 3 kW.

4.1 Datasets of the ICE Vehicles

The employed datasets in the modeling algorithm have been gathered using questionnaires filled out by the randomly selected owners of the commuter light duty conventional ICE vehicles in Tehran. The owners were asked to give merely the commuting data. The datasets include home arrival time ($\mathbf{a}_t$), daily travelled distance ($\mathbf{tr}_d$), and home departure time ($\mathbf{d}_t$) of the vehicles during weekdays. The datasets are illustrated in Fig. 3.

Due to the fact that the Gaussian (normal) PDF is a straightforward distribution, it has been employed in most of the recent studies. However, it usually did not match with the datasets which, in turn, led to inaccurate results. Here, it is suggested to fit appropriate non-Gaussian PDFs to the three mentioned RVs. In fact, a number of PDFs are tested on them and then, the best is selected. As may be seen in Fig. 3a, the Weibull PDF ($f_{\mathbf{d}_t}(t)$) is suggested as the most appropriate function to be fitted to the departure time RV as:

$$f_{\mathbf{d}_t}(t) = \frac{\beta}{\alpha}\left(\frac{t}{\alpha}\right)^{(\beta-1)} e^{-\left(\frac{t}{\alpha}\right)^{\beta}}, t > 0 \tag{9}$$

To model the daily travelled distance, a type III generalized expected value (Gev) PDF is derived as (10). The result is illustrated in Fig. 3b.

$$f_{\mathbf{tr}_d}(d) = \frac{1}{\sigma_{\mathbf{tr}_d}}\left(1 + k_{\mathbf{tr}_d}\frac{\left(d - \mu_{\mathbf{tr}_d}\right)}{\sigma_{\mathbf{tr}_d}}\right)^{-\left(1+\frac{1}{k_{\mathbf{tr}_d}}\right)} e^{-\left(1+k_{\mathbf{tr}_d}\frac{\left(d-\mu_{\mathbf{tr}_d}\right)}{\sigma_{\mathbf{tr}_d}}\right)^{-\frac{1}{k_{\mathbf{tr}_d}}}} \tag{10}$$

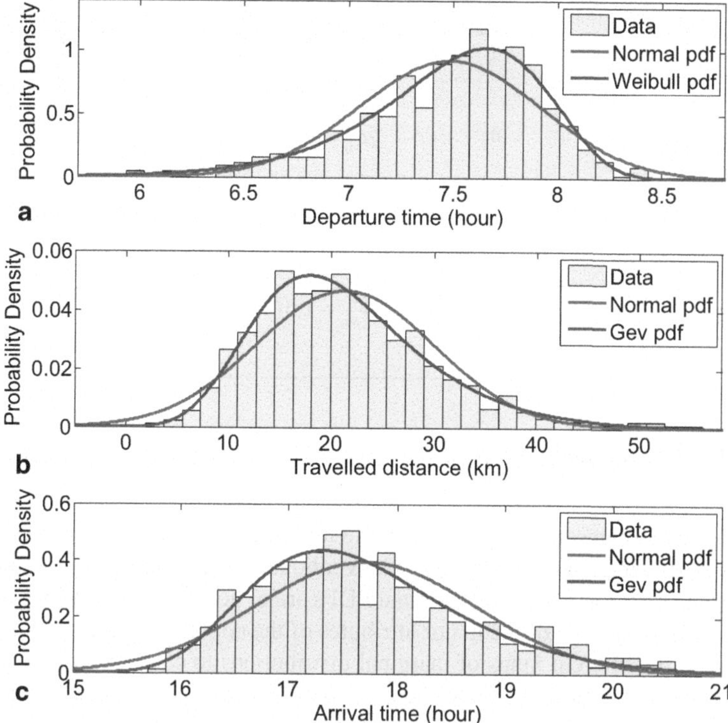

Fig. 3 The data and PDFs of **a** home departure time $\mathbf{d}_t$, **b** daily travelled distance $\mathbf{tr}_d$ and **c** home arrival time $\mathbf{a}_t$

As shown in Fig. 3c, a type III generalized expected value (Gev) PDF ((11)) is fitted to the home arrival time RV as well.

$$f_{\mathbf{a}_t}(t) = \frac{1}{\sigma_{\mathbf{a}_t}}\left(1 + k_{\mathbf{a}_t}\frac{(t-\mu_{\mathbf{a}_t})}{\sigma_{\mathbf{a}_t}}\right)^{-\left(1+\frac{1}{k_{\mathbf{a}_t}}\right)} e^{-\left(1+k_{\mathbf{a}_t}\frac{(t-\mu_{\mathbf{a}_t})}{\sigma_{\mathbf{a}_t}}\right)^{-\frac{1}{k_{\mathbf{a}_t}}}} \tag{11}$$

Further, the extracted distribution of each RV is compared with the normal PDF fitted to the same dataset in order to verify the mentioned claim (Fig. 3). It is seen that the fitted non-Gaussian PDFs provide a better approximation of the original datasets. The parameters of the fitted PDFs are given in Table 2. In the next section, the obtained PDFs are utilized as the marginal distributions to derive a multivariate PDF.

Table 2 The parameters of the fitted PDFs

Datasets	The normal PDF	The suggested PDF
d_t	$\mu_{Nd_t} = 7.48436$	$\alpha = 7.67454$
	$\sigma_{Nd_t} = 0.43178$	$\beta = 21.3812$
tr_d	$\mu_{Ntr_d} = 21.4150$	$k_{tr_d} = \mu 0.052368$
	$\sigma_{Ntr_d} = 8.58711$	$\mu_{tr_d} = 17.6568$
		$\sigma_{tr_d} = 7.1222$
a_t	$\mu_{Na_t} = 17.7170$	$k_{a_t} = \mu 0.060798$
	$\sigma_{Na_t} = 1.01385$	$\mu_{a_t} = 17.2700$
		$\sigma_{a_t} = 0.84832$

4.2 Multivariate Modeling

To generate random samples required in the Monte Carlo simulation, it is essential to fit a copula function to the datasets. Conventional methods are unable to model a multivariate function where the marginal PDFs are of nonidentical distributions. As already mentioned, one of the major attributes of the copula function is its capability to combine different kinds of marginal distributions to construct a multivariate distribution. As the fitted PDFs to the input datasets are not identical, a student's t copula function with 5 degrees of freedom (t_5), C_t: $[0, 1]^3$–$[0, 1]$, is utilized to relate the univariate marginal distributions, $F_{d_t}(t), F_{tr_d}(tr), F_{a_t}(a)$ to their joint distribution function, $H(d, tr, a)$, as:

$$C_t(F_{d_t}(d), F_{tr_d}(tr), \ldots, F_{a_t}(a), \boldsymbol{\rho}_{3\times 3}) = H(d, tr, a) \tag{12}$$

where $\boldsymbol{\rho}$ is extracted as:

$$\boldsymbol{\rho} = \begin{bmatrix} 1 & -0.48 & -0.31 \\ -0.48 & 1 & 0.39 \\ -0.31 & 0.39 & 1 \end{bmatrix} \tag{13}$$

The correlations measure the degree to which large or small values of one random variable associate with large or small values of another.

Following the algorithm described in Sect. 2, the student's t copula function is fitted to the input datasets in order to generate the random samples required in the Monte Carlo simulation (Fig. 4).

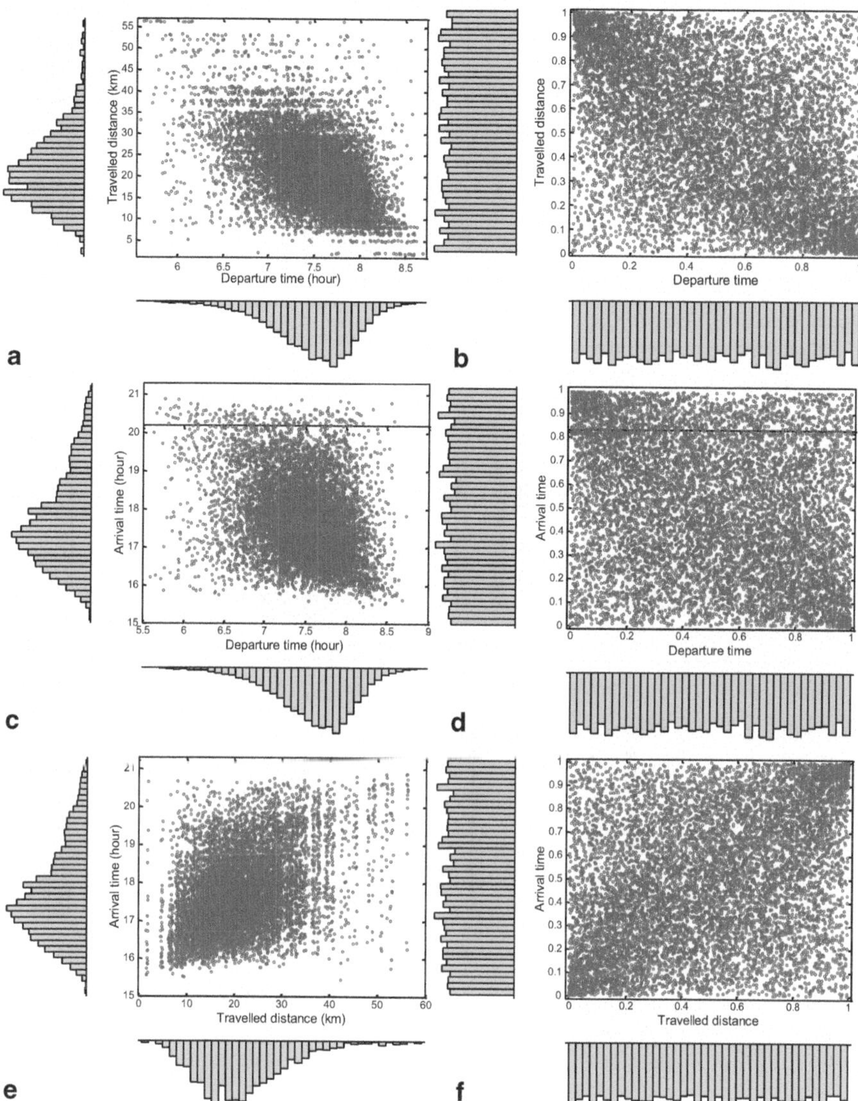

Fig. 4 Scatter plots between departure time and travelled distance (**a, b**), between departure time and arrival time (**c, d**), and between travelled distance and arrival time (**e, f**)

4.3 Synthetic Datasets

Figure 4 illustrates scatter plots of the randomly generated datasets as well as the corresponding distributions in the unit interval. For the sake of the limited space, the plots are merely related to the PEV No. 1. However, the datasets concerned with all the PEVs have almost identical characteristics. Each generated set includes 10,000

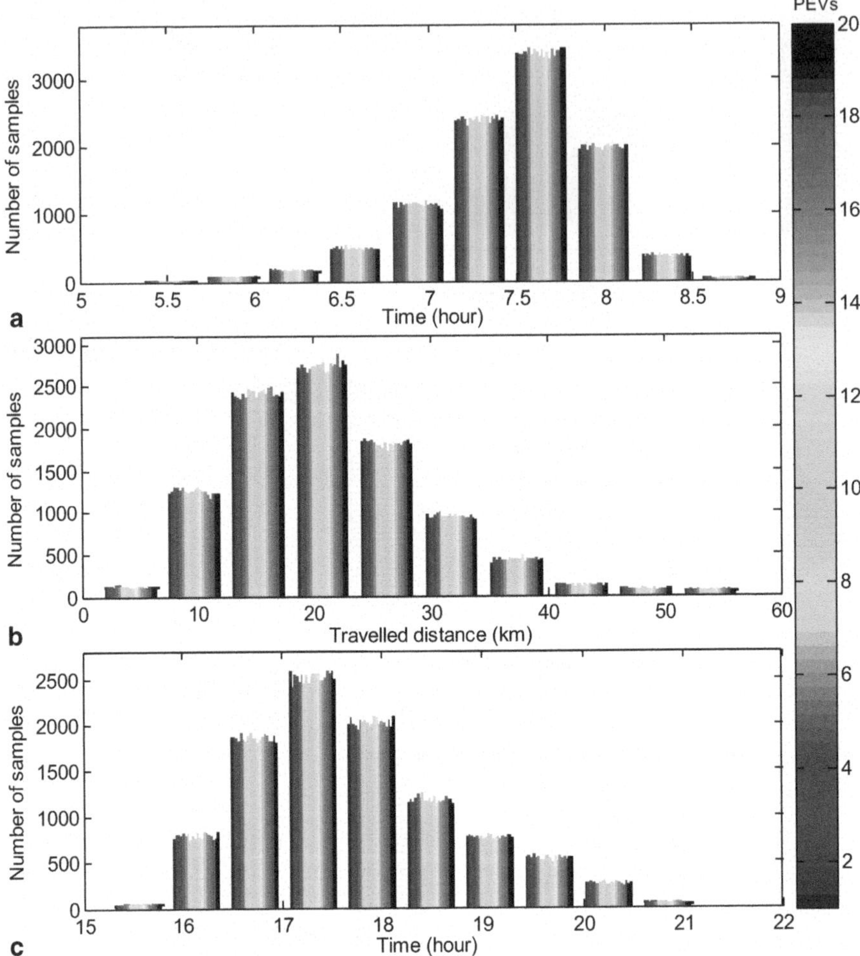

Fig. 5 The randomly generated samples of **a** home departure time $\mathbf{d}_t$, **b** daily travelled distance $\mathbf{tr}_d$, and **c** home arrival time $\mathbf{a}_t$

samples. As is obvious from the histograms, the marginal distributions of the generated samples comply with their original datasets. It is clear that there is a tendency for large (small) values of $\mathbf{d}_t$ to be associated with small (large) values of both $\mathbf{tr}_d$ and $\mathbf{a}_t$, as expected according to (13). Moreover, there is a tendency for large values of $\mathbf{tr}_d$ to be associated with large values of $\mathbf{a}_t$ and similarly for small values. Figure 5 presents distribution of the randomly generated samples (the synthetic data) for all the 20 PEVs. As is obvious from Fig. 5a, the generated samples of $\mathbf{d}_t$ are distributed between 5:20 a.m. and 8:50 a.m. with the highest number of samples between 7:30 a.m. and 8 a.m.. Figure 5b presents the generated samples of $\mathbf{tr}_d$ distributed between 0 and 57 km. The zero distance (0 km) means the PEV departure has not happened. Fig-

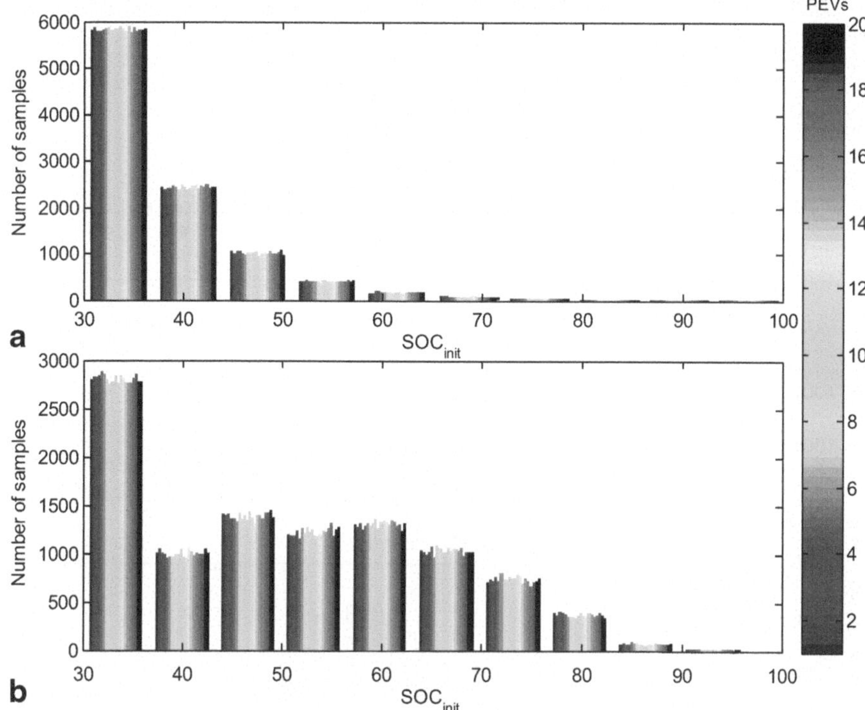

Fig. 6 Battery initial SOC generated by using **a** scenario no. 1, **b** scenario no. 2

ure 5c clearly shows the generated samples of $\mathbf{a}_t$ distributed between 15:15 p.m. and 21:10 p.m. The highest number of samples appears between 17 p.m. and 17:40 p.m.

4.4 The Initial SOC of the Batteries

To derive the initial SOC of the PEV batteries the mentioned scenarios are employed:

Case 1 In (7), f_{exp}(x) is utilized to generate 10,000 random samples. Then, the SOC_{init} is determined accordingly which can be considered as the worst case scenario. The synthetic results can be seen in Fig. 6a. As is obvious, a number of the PEVs have fully charged batteries (SOC_{initn} = 100 %) that may be considered as the untraveled vehicles.

Case 2 The SOC_{init} is determined based on the daily travelled distances. According to (8), the SOC_{initn} depends on the travelled distance (tr_{dn}), the battery capacity (Cap_{bat}), and the efficiency coefficient of PEVs during driving (C_{eff}). Histogram of the synthetic initial SOCs of the PEVs is shown in Fig. 6b.

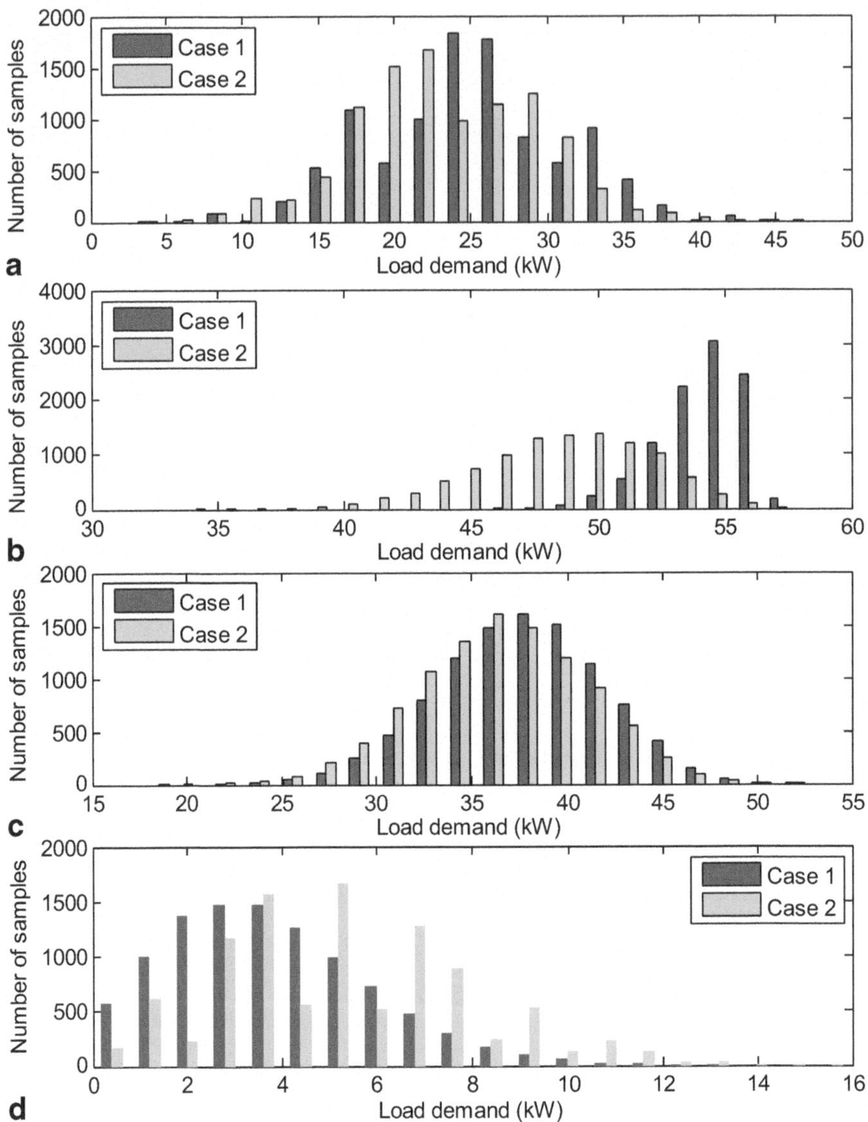

Fig. 7 The distribution of the aggregated power consumption of the PEVs ($DAPF_h$) with respect to two scenarios of the battery initial SOC at **a** h=16 o'clock **b** h=19 o'clock, **c** h=22 o'clock, and **d** h=1 o'clock

4.5 The Aggregated Power Consumption

The demand profile of the fleet (DPF_m) is achieved by estimating the demand profile of the individual PEVs ($DPIP_n$) in each iteration of the devised procedure. As

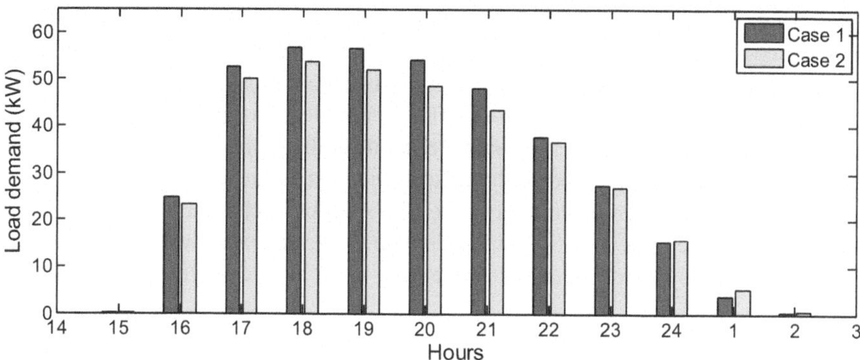

Fig. 8 Demand profile of the fleet (DPF) delivered through a distribution transformer

mentioned in Sect. 3, the estimated profile can be adopted as the profile of the power delivered to the PEVs through the distribution transformer.

According to Fig. 2, the procedure is repeated for 10,000 times in order to derive the distribution of the aggregated power consumption of the fleet within each hour ($DAPF_h$). The estimated $DAPF_h$ at h = 16 o'clock, h = 19 o'clock, h = 22 o'clock, and h = 1 o'clock with respect to two scenarios of the battery initial SOC are shown in Figs. 7a, b, c, and d respectively. These DAPFs can be utilized in order to provide required data in probabilistic planning issues. Moreover, the demand profile of the fleet (DPF) consisted by the expected values of the $DAPF_h$ is presented in Fig. 8.

As may be expected, in most cases, the DPF values calculated by adopting the first scenario (case1) are greater in comparison with the estimated results employing the second scenario (case 2). As an acceptable explanation to this phenomenon, the number of the samples with lower SOC_{init} in case 1 is noticeably higher in comparison to case 2.

In spite of differences among the mentioned scenarios, the resultant DPFs last till the similar morning hours. In order to explain the reason, the samples with late arrivals should be investigated accurately. Due to the correlation structure among the datasets, the mentioned PEVs have relatively lower initial SOCs. Charging occurrence of these PEVs leads to lasting of the DPFs till the similar morning hours.

5 Conclusions

In this contribution, a multivariate stochastic method based on the notion of copula to provide a set of models in order to extract the hourly aggregated load demand of PEVs delivered through a distribution transformer has been developed. A series of datasets, including home arrival time, daily travelled distance, and home departure time of randomly selected private conventional ICE vehicles have been employed. To avoid any mismatch between the original dataset and their probabilistic models, appropriate non-Gaussian probability density functions have been fitted to them. Then, a copula-based multivariate distribution has been constructed by specifying marginal univariate distributions, and afterwards, combining the univariate distributions to provide dependence structure among variables. Extraction of the charging profile has been performed within each iteration for the individual PEVs in order to estimate the hourly aggregated load profile of the fleet. Afterwards, probability density function of the aggregated load of the PEVs within each hour has been derived by applying the Monte Carlo simulation. The expected value of the hourly load demand has finally been calculated concerning the achieved power distributions.

References

1. Tuttle DP, Baldick R (2012) The evolution of plug-in electric vehicle-grid interactions. IEEE Trans Smart Grid 3(1):500–505
2. Golkar MA, Pashajavid E (2011) Analytical assess-ment of mutual impacts between PHEVs and power grid, 21st International Conference on Electricity Dis-tribution (CIRED), Frankfurt, 6–9 June 2011
3. Department of energy, United States of America. One million electric vehicles by 2015. Status report, Feb 2011
4. Saber AY, Venayagamoorthy GK (2009) One million plug-in electric vehicles on the road by 2015. In: Pro-ceedings of the 12th International Conference on In-telligent Transportation Systems, 2009, pp 141–147
5. Yuchao M, Houghton T, Cruden A, Infield D (2012) Modeling the benefits of vehicle-to-grid technology to a power system. IEEE Trans Power Sys 27(2):1012–1020
6. Emadi A, Lee YJ, Rajashekara K (2008) Power elec-tronics and motor drives in electric, hybrid electric, and plug-in hybrid electric vehicles. IEEE Trans Ind Electron 55(6):2237–2245
7. Hajimiragha AH, Canizares CA, Fowler MW, Moazeni S, Elkamel A (2011) A robust optimization approach for planning the transition to plug-in hybrid electric vehicles. IEEE Trans Power Sys 26(4):2264–2274
8. Ferdowsi M (2007) Plug-in hybrid vehicles—a vision for the future. In: IEEE Vehicle Power and Propulsion Conference, VPPC, pp 457–462, 2007
9. Wu D, Aliprantis DC, Gkritza K (2011) Electric energy and power consumption by light-duty plug-in electric vehicles. IEEE Trans Power Sys 26(2):738–746
10. Trovao L, Jorge HM (2011) Power demand impacts of the charging of electric vehicles on the power distribution network in a residential area. In: Proceedings of the 2011 3rd International Youth Conference on Energetics (IYCE), 7–9 July 2011
11. Qian K, Zhou C, Allan M, Yuan Y (2010) Modeling of load demand due to EV battery charging in distribution systems. IEEE Trans Power Sys 26(2):802–810
12. Durante F, Sempi C (2010) Copula theory: an introduction. In: Lecture notes in statistics—proceedings, workshop on copula theory and its applications. Springer, Berlin, pp 3–31
13. Gill S, Stephen B, Galloway S (2012) Wind turbine condition assessment through power curve copula modeling. IEEE Trans Sustain Energy 3(1):94–101
14. Papaefthymiou G, Kurowicka D (2009) Using copulas for modeling stochastic dependence in power system uncertainty analysis. IEEE Trans Power Sys 24(1):40–49
15. Stephen B, Galloway SJ, McMillan D, Hill DC, Infield DG (2011) A copula model of wind turbine perfor-mance. IEEE Trans Power Syst 26(2):965–966
16. Gatz JH (2007) Properties and applications of the student t copula. Master dissertation, Delft Univ. Tech., Jul. 2007. http://risk2.ewi.tudelft.nl/research-and-publications/doc_download/153-gatzthesispdf. Accessed 2012

Probabilistic Home Load Controlling Considering Plug-in Hybrid Electric Vehicle Uncertainties

Mahmud Fotuhi-Firuzabad, Mohammad Rastegar, Amir Safdarian and Farrokh Aminifar

1 Introduction

In the context of energy efficiency, home load controlling (HLC) problem is designated to schedule the operation periods of responsive appliances subjected to practical constraints. HLC problem is usually solved from the perspective of an end-user consumer by either an energy management system (EMS) [1] or a home controller (HC) [2, 3], subjected to constraints declared by the consumer. HLC problem has a variety of input parameters including the electricity tariff, fixed and schedulable consumptions, and possible energy storage system specifications [1]. Plug-in hybrid electric vehicle (PHEVs) technology is an emerging paradigm and a promising solution to tackle the threatening environmental challenges in urban area as [4]. PHEVs, with large batteries connected to the grid to charge [5, 6], could basically work as a distributed energy storage infrastructure for the grid. Accordingly, returning the stored energy back to the grid is offered by the users, if feasible and profitable. PHEVs are usually plugged into charge just after getting back to the home, i.e., more often between hours 18 and 21 [5]. This interval usually coincides with the residential peak consumption when the electricity price is rather high and the distribution systems are heavily loaded. Therefore, PHEVs should be coordinated with HLC programs to avoid disadvantages of their widespread use.

M. Fotuhi-Firuzabad (✉) · M. Rastegar · A. Safdarian · F. Aminifar
Center of Excellence in Power System Management and Control, Department of Electrical Engineering, Sharif University of Technology, Tehran, Iran
e-mail: fotuhi@sharif.edu

M. Rastegar
e-mail: rastegar_m@ee.sharif.edu

A. Safdarian
e-mail: a_safdarian@ee.sharif.edu

F. Aminifar
School of Electrical and Computer Engineering, College of Engineering, University of Tehran, Tehran, Iran
e-mail: faminifar@ut.ac.ir

R. Karki et al. (eds.), *Reliability Modeling and Analysis of Smart Power Systems*, Reliable and Sustainable Electric Power and Energy Systems Management,
DOI 10.1007/978-81-322-1798-5_8,

In recent years, different researchers have concentrated on HLC programming, considering various facilities, and household systems. Reference [1] devises an energy scheduler in an automated home to achieve a desired trade-off between minimizing the electricity payment and minimizing the waiting time for the operation of each appliance. In [2], acceptable bounds on the temperature are used as the operational constraint in the scheduling process to meet the required comfort of the consumer. A decision support tool for smart homes has been proposed in [3] in which a PHEV, a space heater, a water heater, a pool pump, and a photo voltaic (PV) system are scheduled based on various time-of-use (TOU) tariffs. Available distributed energy resources (DER) are then programmed to maximize the net benefit. References [7] and [8] introduce the performance of an in-home energy management (iHEM) application. This application gives more flexibility in using local resources and increases the consumer comfort by interacting with them. Majority of existing works define the problem in a deterministic manner while some of the input parameters have inherent uncertainties. Incorporating these uncertainties would obviously result in more realistic outcomes. Only few works have so far appeared in this regard. References [9] and [10] describe an optimization model to adjust the hourly load level of a given consumer in response to the variable electricity prices. In [9], the price uncertainty is modeled through robust optimization technique.

In a smart home, appliances are either nonresponsive with fixed consumption or schedulable and responsive to the time-varying prices. Demand of the former class is forecasted for the scheduling periods ahead and the latter class defines a set of decision variables to the HLC. Some associated features of the PHEV such as departure time from home, miles daily driven, and energy consumption are exposed to an inherent uncertainty; hence, they should be modeled by appropriate random variables.

This chapter provides a scenario based probabilistic HLC problem that allows a consumer to adapt household facilities in response to the TOU tariff declared for the next day. The present work focuses on the uncertainties of PHEV characteristics, i.e., departure time, travelling time, and out-of-home energy consumption. Not much research work has been reported to incorporate PHEV uncertainties into HLC programming.

The rest of the chapter is outlined as follows. A summary of preliminary subjects around data flowing structure and dynamic pricing is first provided in the next section. Then, PHEVs and their probabilistic nature are characterized and the formulation of the probabilistic HLC problem is developed. The simulation results are presented and a various set of sensitivity studies are conducted. The conclusions are finally drawn in the last section.

2 System Data Structure and Pricing Mechanism

Demand response (DR) as an underlying program at the distribution level allows retail consumers to influence the electricity markets by having the ability to respond to prices as they change over the time. Guiding a way for flowing data is a principle

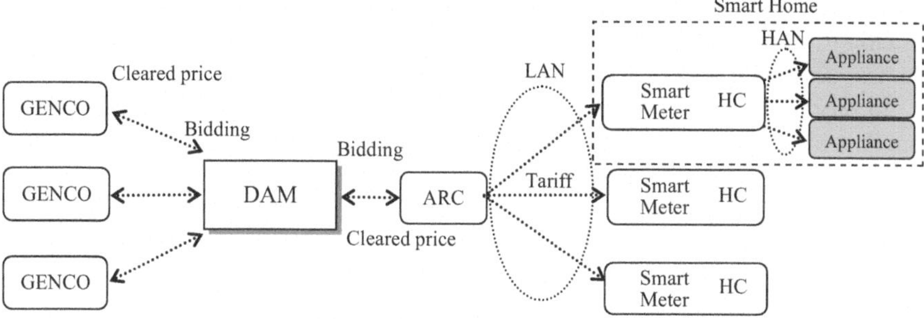

Fig. 1 System data transfer structure

point for DR. In this section, first, we review the data structure utilized in smart grids; thereafter, the dynamic pricing methods, specifically TOU pricing, are briefly discussed.

2.1 Data Structure

The data transfer structure between the market and other entities is depicted in Fig. 1. Generation companies (GENCO) and aggregator of retail consumers (ARC) bid their offers to the day ahead market (DAM), based on their associated forecasting load and bidding strategies. The market is then cleared by the ISO and prices are announced to the market participants. Next, the ARC declares retail prices for end-users mainly based on the whole sale market prices. In a home, the smart meter receives the price data via a local area network (LAN) [9]. The HC dedicated to each smart meter is responsible to control loads by setting up operation times of appliances. The HC solves the HLC problem taking time-varying tariffs into account. As shown in Fig. 1, control signals are transmitted from the HC to responsive appliances over a home area network (HAN) [9].

2.2 Dynamic Pricing

Time differentiated pricing models can potentially lead to economic and environmental advantages compared to the current common flat rates [10]. As noted earlier, they can, in essence, provide for end-users the opportunity of reducing their electricity expenditures by responding to the price that varies during different periods of a day.

The time differentiated pricing is currently implemented in several utilities. The basic idea behind the time-varying pricing is to incense users to respond to the price. Encouraging users to shift high-load household appliances to off-peak hours not only reduces their electricity costs but also declines the peak to average ratio (PAR) in the load profile. For this purpose, various time-differentiated pricing models have been proposed as yet, including: real-time pricing (RTP), TOU pricing, critical peak pricing (CPP), etc., [2, 11]. ARC can apply each of these models to its served consumers according to the agreement.

In recent years, TOU energy prices are researched by many scholars and also implemented by a large number of utilities [12]. It is recognized that carrying out TOU energy prices has significant impacts on the load shifting [11]. This pricing mechanism is based on the fact that the energy system's operation cost in the time of peak-load is much higher than that in the valley time [13]. The above principle insures that the energy price is higher when the consumption is higher. With TOU prices, the price of electricity is time dependent. When demand is low, less expensive sources of electricity are used. When demand rises, more expensive forms of electricity production are called upon, making prices higher. In this chapter, a three-level (on-peak, mid-peak, off-peak) TOU tariff is utilized for the HLC problem.

3 PHEV Characteristics

New generation of vehicles such as PHEVs, which have an electric motor along with an internal combustion engine, are recently becoming more popular and would broadly be seen on the roads in the near future. They are referred to as vehicles rechargeable by connecting to an external electric power source. These new electrical loads could have negative impacts on distribution grids. On the contrary, at peak hours, PHEVs, by returning electricity back to the grid, can alleviate the network congestions and improve the system reliability. This capability, generally known as vehicle to grid (V2G) operation mode, incenses the vehicular system communicate with the grid in order to sell back electricity into the grid, if profitable [11].

To overcome the PHEV's disadvantages and promote their benefits, it is necessary to investigate their characteristics thoroughly. PHEV characteristics can be divided into two categories based on their effects on the HLC; the first category includes features related to PHEV and the battery type such as battery capacity and charging/discharging rate. The second class comprises properties that depend on the travelling habits of the vehicle owner. The former category is deterministic and does not change much over the time. However, detailed information about vehicle trips is currently unavailable for the consumer due to the lack of comprehensive study on travelling habits in previous literatures [4]. Accordingly, the consumer just estimates the second category of PHEV parameters for the next day. These estimates suffer from a significant level of uncertainties. The uncertainty should be incorporated in models for the sake of achieving more closeness between the model and actual conditions. Otherwise, HLC might result in non-optimal solutions from the economical perspectives.

Probabilistic models have been identified as a tool to analyze random behavior of the system. They are strongly based on the historical data. Reported data in [12] has suggested that the uncertainty can be reasonably described by a normal distribution, shown in Fig. 2. To this end, the probability distribution parameters, i.e., mean

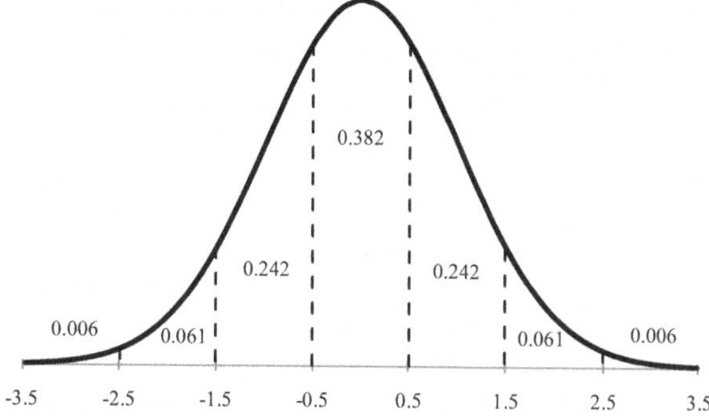

Fig. 2 Seven-step approximation of the standard normal distribution with unit standard deviation

and standard deviation, can be determined from the consumer past experiences. When sufficient information is not available, the forecasted value is a proper option for the mean value. The standard deviation can also be treated according to the vehicle owner behavior.

After selecting the appropriate values for the mean and standard deviation, there is also another difficulty. It is difficult to include such a continuous probability density function (PDF) into the mathematical framework. In such a case, it is a common approach to use Monte Carlo simulation method for generating appropriate number of scenarios. However, to achieve a reasonable level of accuracy, production of a large number of scenarios is required which in turn could be computationally expensive. To avoid such a complexity, the PDF can be divided into a discrete number of class intervals. In such a case, the area of each interval represents the probability of the class interval midpoint. Figure 2 demonstrates a normal PDF which is divided into seven intervals [13].

Ultimately, it should be noted that a similar procedure is used to model the uncertainty of the departure time, consumed energy, and travelling time. Also, these parameters are assumed to be independent and therefore, the occurrence probability of each scenario is the product of occurrence probabilities associated with the individual elements.

4 HLC Framework and Formulation

One of the most important features of MIP method is that the MIP solvers are capable of ensuring that the obtained solution is global optimal or one within an acceptable tolerance. Moreover, MIP method is able to directly measure the optimality of a solution. In this section, HLC framework is defined to schedule the decision variables, i.e., operation periods of the responsive appliances and charging/discharging cycles of PHEV battery, based on the MIP method. The objective function is to minimize the cost of providing different appliances including responsive loads, nonresponsive loads, and PHEV within the upcoming scheduling horizon. The objective function is expressed as follows:

$$C = \sum_{h \in [1,\ldots,NP]} C_h^{NR} + C_h^{R} + C_h^{PHEV}, \tag{1}$$

where, NP is the number of next schedulable periods and h is the index of period. The first term of the objective function denotes the cost of supplying nonresponsive loads at period h. In this chapter, the amount of such loads is assumed to be known at each hour of the day. Hence, the first term of Eq. (1) is a constant value and can be omitted from the model. The second term of the objective function designates the cost of responsive loads operation. This group of loads have definite amount of power consumption while their operation periods should be scheduled by solution procedure. The third term, as the cost associated with PHEV, includes the positive and negative costs of PHEV charging and discharging processes, respectively, as well as the cost of out-of-home gasoline consumption. This term is a function of charging and discharging cycles of the PHEV, which are among the problem decision variables. In this chapter, charge and discharge rates of the PHEV battery are assumed to be known. The departure time, the amount of out-of-home energy consumption, and travelling time are assumed to be forecasted by the consumer. According to the previous section, uncertainties of these parameters are modeled as independent discrete normal PDFs. To integrate these uncertainties into the model, a set of possible scenarios for the PHEV parameters are constructed. These scenarios consist of departure time, duration of the travelling time, and out-of-home energy consumption.

Incorporating the impact of uncertainty, the cost function should be minimized over the constructed scenarios. Therefore, the objective function will be as

$$C = \sum_{s} pr_s \times \sum_{h} C_h^{NR} + C_h^{R} + C_{s,h}^{PHEV} \tag{2}$$

where, s is the scenario index and pr_s is its associated probability. The cost terms in Eq. (2) are defined as

$$C_h^{NR} = T_h E_h^{NR}, \qquad \forall h, \tag{3}$$

$$C_h^{R} = T_h E_h^{R}, \qquad \forall h, \tag{4}$$

$$C_{s,h}^{PHEV} = T_h E_h^{PHEV} + C_{s,h}^{gas}, \quad \forall h, \forall s, \tag{5}$$

where, E_h^{NR}, E_h^{R}, and E_h^{PHEV} are the energy consumption of nonresponsive appliances, responsive appliances, and PHEV at period h [kWh], respectively. T_h indicates the predetermined tariff of period h. As it can be seen in Eq. (5), the cost associated with PHEV consists of two parts. The first and second parts are related to the cost of charged/discharged energy and the cost of extra gasoline consumption, respectively. It should be noted that battery and combustion engine are working together and the cost of usual gasoline consumption is not considered here. In this chapter, it is assumed that in the case of electrical energy shortage, PHEV will automatically switch onto gasoline. This attribute results in a penalty cost associated with the higher gasoline price.

So far, the objective function of the scheduling problem is formulated. This problem is subject to a set of constraints, which are digested in the following.

Total energy consumption of responsive appliances at period h is formulated as

$$E_h^R = \sum_i b_h^i E^i, \quad \forall h, \tag{6}$$

where, i is the index of responsive appliances, E^i is the energy consumption of appliance i at each period, and b_h^i is a binary variable that indicates the status of ith responsive appliance at period h. Equation (7) forces the problem to energize each responsive appliance within the allowable operation time interval according to the consumer decision.

$$U^i = \sum_{h \in AI^i} b_h^i, \quad \forall i, \tag{7}$$

where, AI^i denotes the allowable interval of the ith appliance operation specified by the consumer. U^i is the predetermined number of operation periods for appliance i. Obviously, AI^i should be selected equal to or greater than U^i. It is usually requested that the appliances operate uninterruptedly. Equation (8) guarantees the consecutive operation of the responsive appliances.

$$\sum_{h=t}^{t+U^i-1} b_h^i \geq U^i SU_t^i, \quad \forall t \leq NP - U^i + 1, \tag{8}$$

where, SU_t^i is the startup indicator of appliance i at period t. Equation (9) is to relate the appliance status, startup indicator, and shutdown indicator at period $h \in \{1, \ldots, NP\}$.

$$SU_h^i - SD_h^i = b_h^i - b_{h-1}^i, \quad \forall h. \tag{9}$$

Since an appliance at a given period cannot simultaneously start up and shut down, we have

$$SU_h^i + SD_h^i \leq 1, \quad \forall h. \tag{10}$$

In the following, the constraints associated to PHEV are presented. The amount of energy exchanged bi-directionally between the grid and PHEV battery at period h is formulated mathematically as

$$E_h^{PHEV} = \frac{1}{\eta}(E^{ch} b_h^{ch}) - \eta(E^{dch} b_h^{dch}), \quad \forall h, \tag{11}$$

where, E^{ch} and E^{dch} represent the charged and discharged chemical energies in each period, respectively. In this formula, the energy conversion efficiency indicated by η in both directions, i.e., DC/AC and AC/DC, is assumed the same. Also, b_h^{ch} and b_h^{dch} are binary variables indicating charging and discharging statuses of the battery at period *h*, respectively. It is evident that the battery could not charge and discharge simultaneously. Hence

$$b_h^{ch} + b_h^{dch} \leq 1, \quad \forall h. \tag{12}$$

The other point is that PHEV can be supplied from both the battery storage and gasoline energy. Synonymously, gasoline provides the difference of the out-of-home energy required and the received energy from the battery. So,

$$C_{s,h}^{gas} = \lambda(E_{s,h}^{out} - E_{s,h}^{bat}), \quad \forall h, \forall s. \tag{13}$$

where, $C_{s,h}^{gas}$, $E_{s,h}^{out}$, and $E_{s,h}^{bat}$ are the gasoline consumption cost, the out-of-home energy consumption, and the provided out-of-home energy by the battery at period h in scenario *s*, respectively. λ is the gasoline price [$¢/kWh$]. Evidently, $E_{s,h}^{out} - E_{s,h}^{bat}$ is positive and total energy provided by the battery in the out-of-home interval should be less than PHEV's charge level at the departure time. This condition is satisfied by

$$\sum_{h \in [g,c]} E_{s,h}^{bat} \leq E_0^{bat} + \sum_{h \in [1,g)} (E^{ch} b_h^{ch} - E^{dch} b_h^{dch}) \tag{14}$$

where, E_0^{bat} is the preliminary charging level of PHEV battery, and $[g, c]$ is the PHEV out-of-home interval. On the other hand, the energy stored in the battery should rationally be positive and less than the capacity at each period. Such a constraint could be expressed as

$$0 \leq E_0^{bat} + \sum_{t \in 1,h} (E^{ch} b_t^{ch} - E^{dch} b_t^{dch} - E_{s,t}^{bat}) \leq E^{cap}, \quad \forall h, \tag{15}$$

where, E^{cap} is the capacity of PHEV battery. The discharging value at each period should reasonably be less than the available battery charge. Hence

$$E^{dch} b_h^{dch} \leq E_0^{bat} + \sum\nolimits_{t \in 1,h)} (E^{ch} b_t^{ch} - E^{dch} b_t^{dch} - E_{s,t}^{bat}), \quad \forall h. \tag{16}$$

Due to technical specifications, the energy streamed from the utility toward the consumer is capped by a maximum level, $E^{\max}$, as Eq. (17). The level may be set by the utility to prevent the distribution system congestion. That is,

$$E_h^R + E_h^{NR} + E_h^{PHEV} \leq E^{\max}, \quad \forall h. \tag{17}$$

In case the discharged energy is more than the home-own consumption, the energy could be sold to the utility. Here, the price of selling energy is assumed to be the same as the purchase price.

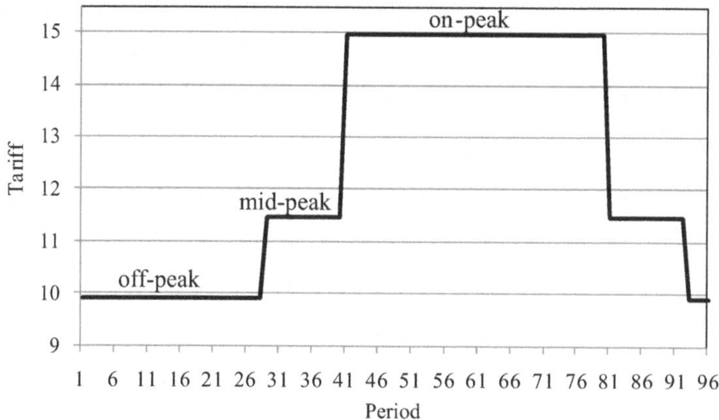

Fig. 3 TOU tariff (¢)

5 Numerical Studies

In this section, the simulation results are presented and the performance of proposed formulation is investigated. As a case study, a single home with a PHEV and some responsive and nonresponsive appliances is considered. The simulations mainly focus on the scheduling of these facilities during the next day with the time steps of 15 min. The energy streamed from the utility toward the consumer is limited to 1.375 kWh/period. Figure 3 shows the three-level tariff taken from [14]. According to this figure, on-peak tariff, between 10 a.m. and 8 p.m. is 14.958 ¢/kWh; mid-peak tariff between 7 a.m. and 10 a.m. and between 8 p.m. and 11 p.m., is equal to 11.453 ¢/kWh; in the remaining periods the off-peak tariff is set equal to 9.866 ¢/kWh.

The given home has two types of typical responsive and nonresponsive household appliances. Nonresponsive load curve forecasted for each hour is drawn in Fig. 4. Also, Table 1 outlines the energy consumption, the number of operation periods (U^i), and the allowable interval (AI^i) for each responsive appliance [15]. All required characteristics of the PHEV's battery are summarized in Table 2. These characteristics include efficiency, capacity, forecasted departure time, forecasted arrival time, and forecasted amount of energy consumption. It is assumed that the battery could totally charge or discharge during 5 h (20 periods).

So, the charged and discharged rates are $10/20 = 0.5$ kWh/period. Also, the preliminary charge level of the battery is assumed to be 1 kWh. Finally, gasoline price is assumed to be 20 ¢/kWh. The following case studies are conducted and the associated HLC results are investigated in depth:

Case I A smart home with the appliances and without PHEV and energy transfer limit.

Case II A smart home with PHEV and energy transfer limit. Parameters of all appliances and PHEV are assumed to be known. Also, V2G is not considered here.

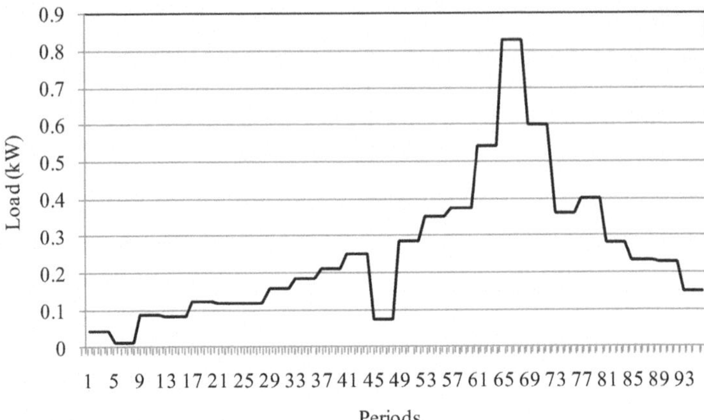

Fig. 4 Forecasted nonresponsive load curve

Table 1 Responsive appliances data

Appliance	E^i (Wh/period)	U^i	AI^i
Clothes dryer	971.25	4	1–96
Dishwasher	300	3	27–52

This indicates that the in-home discharging process of PHEV battery is infeasible. Accordingly, HLC is considered as a deterministic programming.

Case III This case is similar to Case II but with the V2G capability. Synonymously, PHEV can supply other appliances and the network in this case.

Case IV In this case, the uncertainties associated with PHEV parameters are incorporated. However, V2G is not allowed.

Case V In this case, in addition to the uncertainties associated with PHEV parameters, the V2G capability is also joined to the probabilistic problem.

It should be noted that the standard deviation associated with the PHEV uncertain parameters, i.e., departure time, travelling time, and the out-of-home energy consumption, are assumed to be 15 min, 15 min, and 1 kWh, respectively. Seven-step approximation of the normal distribution is used to model the uncertainty of these parameters. Consequently, there are $7 \times 7 \times 7 = 343$ scenarios within the next day horizon with different weights resulted from the multiplication of the parameters occurrence probabilities. The mean of trip departure time, duration time, and energy consumption are assumed period 37, 28 periods, and 8 kWh, respectively, similar to the forecasted values shown in Table 2.

All of the mentioned cases are constructed and the obtained MIP models are solved using CPLEX 11.0.0 under GAMS environment [16]. In the following, the results associated with these cases are discussed.

Table 2 PHEV characteristics

η	E^{cap} (kWh)	g	c	E^{out} (kWh)
98 %	10	37	64	8

Table 3 Operating periods of responsive appliances and consumer payment in cases I-V

Case	Clothes dryer	Dishwasher	Objective function (¢)	Expected payment (¢)
I	1–4	27–29	130.503	130.503
II	3–6	27–29	200.975	210.683
III	25–28	27–29	191.811	202.472
IV	1–4	27–29	208.760	208.760
V	93–96	27–29	198.754	198.754

In Case I, a home with two responsive appliances, scheduled over a 24-h horizon, is studied in depth. This is a simple case in which the correctness of results can be readily verified, yet it conveys many interesting features of the proposed HLC formulation. We derive and assess the effect of scheduling on the PAR of the home load profile. PAR is a key economical indicator reflecting the utilization factor of the investment allocated in all generation, transmission, and distribution hierarchy. Although this concern does not directly belong to the end-users, utility tariff regulations and DR programs implicitly bring about declining PAR. For the sake of comparison, the home energy consumption without any scheduling is obtained. Referring to Fig. 4, the non-responsive peak load is 0.828 kW. Suppose that, clothes dryer is on during peak load periods and dishwasher is on during off-peak tariff periods; thus, the peak consumption grows to 4.712 kW. The average of total energy consumption is 0.1133 kWh/period and PAR = 10.394. Also, the payment cost is equal to ¢ 141.131. The HLC is solved through the proposed formulation. The responsive appliances status and the payment obtained for all the cases can be found in Table 3.

Referring to Table 3, in the aforementioned cases, responsive appliances are on in consecutive periods. Clothes dryer, whose operation interval is not restricted, expectedly lies in the off-peak tariff periods. Also, the maximum number of off-peak periods (periods 27 and 28) lying in the dishwasher's allowable operation interval are exploited. To meet the energy transfer limit, clothes dryer statuses are amended in the various cases.

The results in Case I show that the peak load and PAR are 3.9288 and 8.667 kW, respectively. The consumer payment is meanwhile equal to ¢ 130.503. It should be noted that the average load is the same as what calculated before the scheduling. What makes the results particularly interesting is that, in addition to cost reduction, PAR is descended in comparison with what calculated before. This observation brings out from the similarity of the load and tariff profiles. The consumption at off-peak tariff periods is equivalent to fill valley of load curve. Figure 5 portrays the total load curve.

According to Fig. 5, a concern exists that if all homes in a region run their high consumption appliances during the same periods, the regional demand increases and the distribution network may be congested. One solution to this problem is to

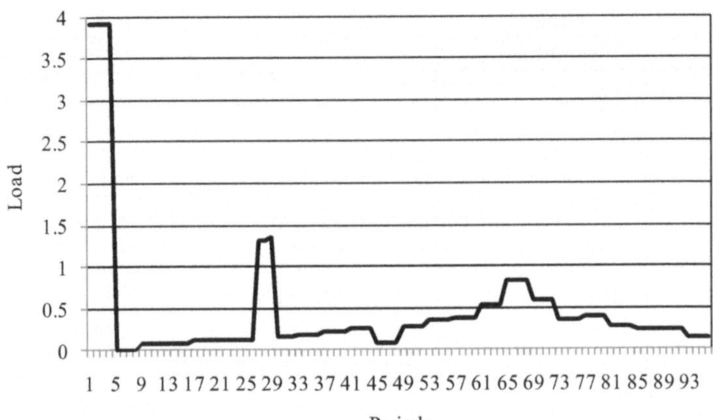

Fig. 5 Load (kW) curve in Case I

limit the energy transfer between homes and the grid which imposes inconvenience to consumers. By a deeper analysis of the results, it can be deduced that the HLC problem does not have a unique solution. As an instance, assume that clothes dryer is on from periods 15 to 18. Tariffs of periods 15–18 are the same as those of operating periods in the studied cases. Therefore, the payment imposed to the consumer remains constant. The attribute of HLC multiple-solution would be profitable for the concurrent scheduling of several homes and avoidance of the network congestion.

It is deduced from Table 3 that in the probabilistic cases, i.e., Cases IV and V, the consumer payments according to the objective function are more than deterministic ones; because, in some scenarios whose out-of-home energy consumption is high, gasoline provides the required energy beyond the battery capacity.

Although a definite appliance and PHEV scheduling is resulted from the cases, the PHEV energy consumption, travelling time, and departure time may alter unexpectedly. Thus, an average cost should be calculated in such modes according to each event probability. This cost is designated as expected payment of the consumer. Since, in probabilistic cases, i.e., Cases IV and V, all the probable scenarios were considered in the scheduling problem, the resulted cost of objective function and expected payment are the same. On the other hand, unexpected variation in PHEV parameters causes more gasoline consumption in deterministic scheduling. So, as found in Table 3, the expected cost is more than the objective function in Cases II and III. Also, the results show that the expected payment in probabilistic cases is lower than deterministic ones.

The PHEV charging and discharging statuses, derived from deterministic HLC solving, are shown in Fig. 6. In this and forward figure, $+1$ and -1 stand for charging and discharging modes, respectively. Expectedly, the charging periods are all among off-peak tariff periods. Figure 6 illustrates that during periods 37 to 64, when the PHEV is unplugged, no charging or discharging takes place. The energy needed

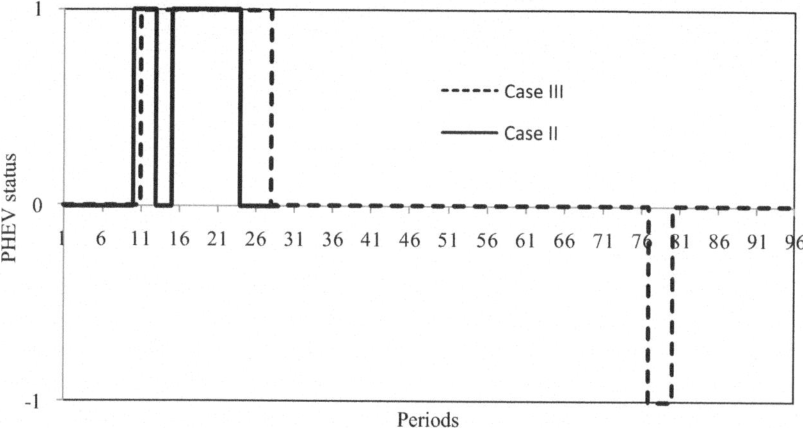

Fig. 6 PHEV status in Cases II and III

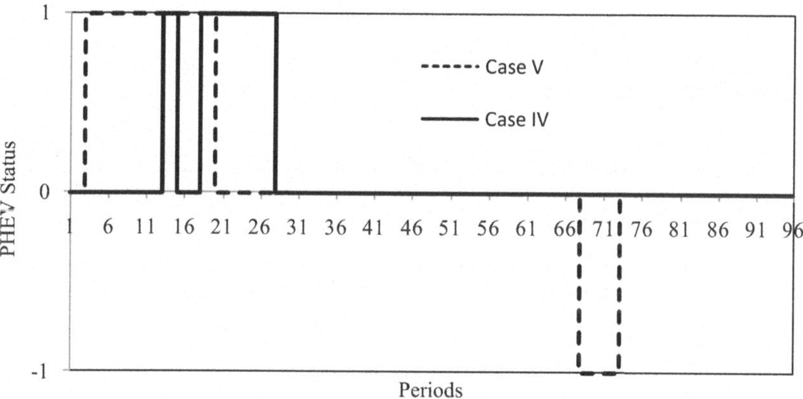

Fig. 7 PHEV status in Cases IV and V

for PHEV outdoor operation is stored before the period 37. The HLC results also represent that the amount of the energy received from the battery in the out-of-home interval equals to the PHEV trip consumption, i.e., 8 kWh. So, in these deterministic cases, due to the higher price of gasoline than the electricity, gasoline is expectedly not utilized during the trip.

Figure 7 portrays the PHEV status resulted from probabilistic HLC solution in Cases IV and V. The charging and discharging periods are entirely changed in Case V in comparison with HLC without discharging capability. Before going out of home (period 37) the PHEV battery is charged during periods 14 and 18 in Cases IV and V, respectively. Although the level of the PHEV battery charge before departure time is more in Case V than that of Case IV, the total gasoline consumption in different scenarios in Case V is also more than that of in Case IV. The reason for this

Table 4 Consumer payment versus different departure time deviation

Standard deviation (min)	Payment (¢)
15	198.754
30	198.754
60	198.754
120	220.874

observation is that PHEV in Case V prefers to store energy and sell it back to the grid during on-tariff periods after coming back home.

Figures 6 and 7 represent that, in Cases III and V, after coming back home (period 64), the battery is discharging 4 and 6 periods, respectively, during on-peak tariff periods. Since the energy which has been stored during off-peak tariff periods is returned to other appliances at on-peak tariff periods and the home does not buy the energy during high-price periods, this scheduling undoubtedly affords some money saving for the consumer. In addition, excessive returned energy is sold to the grid and the consumer makes more money. So, the consumer payments in Cases III and V are lower than that of Cases II and IV, respectively, as presented in Table 3.

6 Sensitivity Analyses

In this section, various amount of standard deviations associated with departure time, travelling time, and out-of-home energy consumption are considered to reveal the effects of these parameters' uncertainty. For the sake of illustration, the last case of the previous section is adopted and the consumer payment resulted from probabilistic HLC solving are presented and discussed in the following. In each subsection, the deviation of the not-mentioned uncertain parameters are assumed the same as Case V.

6.1 Different Levels of Departure Time Uncertainty

The departure time of the PHEV is an effective component in the scheduling problem. The level of the uncertainty associated with this parameter is studied in this subsection. To investigate this issue, the proposed formulation is examined by different levels of departure time uncertainty and the obtained results are outlined in Table 4.

From Table 4, it can be deduced that the payment does not change in the first three levels of departure time uncertainty. This observation is justified knowing that tariffs are constant during charging procedure and also there is enough time for PHEV to become full charged in first three cases. Therefore, payment cost may be different in cases with the earlier PHEV departure time. For example, consider the mean of the departure time be in the early morning so that the PHEV battery has less time to charge. Hence, a small deviation would make PHEV consume further

Table 5 Consumer payment versus different duration time deviation

Standard deviation (min)	Payment (¢)
15	198.754
30	198.754
60	199.590
120	206.444

Table 6 Consumer payment versus different energy consumption deviation

Standard deviation (kWh)	Payment (¢)
0.5	195.350
1	198.754
1.5	202.159
2	206.444

amount of the gasoline during the trip. This imposes higher cost to the consumer. Also, as illustrated in the fourth row of the deviation in Table 4, greater deviations causes more payment due to the above discussed reason.

6.2 *Different Levels of the Travelling Time Uncertainty*

In this subsection, the level of the travelling time uncertainty is investigated. Four cases with different standard deviations are generated and the HLC problem is solved. The obtained results are shown in Table 5. As shown, the results are expectedly analogous to the previous subsection. Fifteen and thirty minute deviation have no impacts on the scheduling. However, 60 and 120 min deviation changes the charging/discharging cycles of the PHEV. A long duration trip increases payment cost in two ways. The amount of required expensive gasoline is more in a long duration trip. Also, the amount of energy discharged into the grid after coming back home becomes lower after a long duration trip. As a result, the higher payment is expected due to the trip duration deviation increment.

6.3 *Different Levels of the Consumed Energy Uncertainty*

In this subsection, the impact of different levels of the consumed energy uncertainty on the solution is probed. Accordingly, various standard deviations of consumed energy are simulated. Table 6 outlines the results. According to the obtained results, by increasing the energy consumption uncertainty level, i.e., standard deviation, payment becomes greater. These results would be the subsequent of either more gasoline consumption or less capability of discharging after coming back home.

7 Conclusion

This chapter proposed a MIP model to economically schedule the consumption periods of responsive appliances and the PHEV battery charge/discharge periods. Uncertain behavior of PHEV characteristics including departure time, travelling time, and energy consumption were also incorporated into the model. Three independent seven-step estimation of the normal PDFs were used to cope with these uncertain parameters. Numerical studies and sensitivity analysis with a large range of cases were conducted in this chapter. Based on the obtained results, V2G operation mode of PHEV can dramatically reduce the providing cost. Also, it is illustrated that probabilistic scheduling will result in a lower expected payment cost than the deterministic scheduling. Moreover, it was deduced from the sensitivity analysis that the increment of PHEV characteristics uncertainties would impose consumer to pay more. The concepts presented in this chapter can be extended to include other uncertainties existing in actual practices.

References

1. Mohsenian-Rad AH, Leon-Garcia A (2010) Optimal residential load control with price prediction in real-time electricity pricing environment. IEEE Trans Smart Grid 1:120–133
2. Du P, Lu N (2011) Appliance commitment for household load scheduling. IEEE Trans Smart Grid 2:411–419
3. Pedrasa MAA, Spooner TD, MacGill IF (2010) Coordinated scheduling of residential distributed energy resources to optimize smart home energy services. IEEE Trans Smart Grid 1:134–142
4. Darabi Z, Ferdowsi M (2011) Aggregated impact of plug-in hybrid electric vehicles on electricity demand profile. IEEE Trans Sustain Energy 2:501–508
5. Clement-Nyns K, Haesen E, Driesen J (2010) The impact of charging plug-in hybrid electric vehicles on a residential distribution grid. IEEE Trans Power Syst 25:371–380
6. Vazquez S, Lukic SM, Galvan E, Franquelo LG, Carrasco JM (2010) Energy storage systems for transport and grid applications. IEEE Trans Indus Electron 57:3881–3895
7. Erol-Kantarci M, Mouftah HT (2011) Wireless sensor networks for cost-efficient residential energy management in the smart grid. IEEE Trans Smart Grid 2:314–325
8. Erol-Kantarci M, Mouftah HT (2010) Using wireless sensor networks for energy-aware homes in smart grids. IEEE Symposium on computers and communications (ISCC)
9. Conejo AJ, Morales JM, Baringo L (2010) Real-time demand response model. IEEE Trans Smart Grid 1:120–133
10. Rastegar M, Fotuhi-Firuzabad M, Aminifar F (2012) Load commitment in a smart home. J Appl Energy 96:45–54
11. Ipakchi A, Albuyeh F (2009) Grid of the future. IEEE Power Energy Mag 7(4):52–62
12. National household travel survey (2014) http://nhts.ornl.gov
13. Billinton R, Allen R (1996) Reliability evaluation of power system. Plenum Press
14. Baltimore gas and electric three-level summer's tariffs (2012) http://www.bge.com/portal/site/bge/menuitem.dc72e1697738765822b75475da6176a0/
15. The infinite power of Texas, estimating PV system size and cost. SECO fact sheet, vol 24, pp 1–4, 2008
16. Brooke A, Kendrick D, Meeraus A, Raman R (2003) GAMS: a user's guide. GAMS Development Corp., Washington, DC

A Load Management Perspective of the Smart Grid: Simple and Effective Tools to Enhance Reliability

Amir Moshari and Akbar Ebrahimi

1 Introduction

The electric power system is experiencing an important change due to the emerging concerns, such as the energy conservation, environmental considerations, better asset utilization, improved system reliability, higher quality of service, enhanced customer choice, etc. These considerations will widely change the electricity distribution grid because it is not observable; it employs electromechanical components; it is almost operated manually; and in short, it needs to be transformed into a "smart grid." This transformation will be essential to meet new environmental requirements, to support plug-in hybrid electric vehicles (PHEVs), storage capabilities and distributed generation, and to improve demand response (DR) [1]. In brief, "a smart grid is the use of sensors, communications, computational ability, and control in some form to enhance the overall functionality of the electric power delivery system" [2].

Reliability has always been an essential issue in power systems design and operation. In the USA, it is estimated that the annual cost of outages was about \$ 79 billions in 2002, which is almost equal to a third of the total electricity retail revenue (\$ 249 billions) [3]. However, meeting reliability targets is going to be more difficult in modern power grids due to various factors [3]:

- Electricity markets increase the number of transmission and the amount of transmitted energy over long distances, which will result in aggravated volatility and reduced reliability margins.
- Uncertainty, distribution, and diversity of energy supplies because of environmental sustainability concerns, lead to aggravated grid congestion. Such situation

A. Moshari (✉) · A. Ebrahimi
Department of Electrical and Computer Engineering,
Isfahan University of Technology, Isfahan, Iran
e-mail: a.moshari@ec.iut.ac.ir

A. Ebrahimi
e-mail: ebrahimi@cc.iut.ac.ir

R. Karki et al. (eds.), *Reliability Modeling and Analysis of Smart Power Systems*, Reliable and Sustainable Electric Power and Energy Systems Management,
DOI 10.1007/978-81-322-1798-5_9, © Springer India 2014

will result in some real-time power flow patterns which are quite different from those that offline analyses show.

- Asset utilization schemes force the grid to be operated at its "edge" more often and in more places.
- Wide implementation of distributed resources diminishes the distinction between transmission and distribution parts of power systems, and will increase volatility and complexity of the power grid.
- Combination of operating entities leads to more complicated problems along with shorter times for deciding about them and smaller margins for errors.

Load management schemes have been employed since the early 1980s. Peak clipping, valley filling, load shifting, energy conservation, direct load control (DLC), and other load management programs have been implemented by many utilities with different degrees of success. Now, due to increasing interests in demand-side management and energy conservation (because of the environmental concerns), DR is going to play a more serious role in the control and management of power systems. Also, the advanced metering infrastructures (AMI), which can be employed in the future smart grids, will facilitate the implementation of dynamic tariffs, participation of retail demand-side into the wholesale energy markets, and management and control of demand-side energy resources. In the future smart grids, demand-side resources can be managed in order to meet the available generation at any time, considering the power grid delivery capabilities. Therefore, by emerging smart grids, the present load-following operating strategy will move toward a load-shaping strategy [1].

Most of the researches in smart grid area are still in the stage of qualitative definitions and development of concepts. Much of the existing references have mainly described definitions, characteristics, and elements of the future smart grids, and the benefits and features which will be provided by smart grids [1, 4–7]. However, some references have also mentioned the smart grid reliability and the effects of future technologies on system reliability. Among them, reference [3] has reviewed the reliability perspectives of the smart grid and the effects of the smart grid's sources like renewable resources, DR, storage devices, and electric transportation on system reliability. Other references have mainly discussed about issues like load control (LC) and DR [8–11], distributed generation [12, 13], electric vehicles and storage [1, 14], control, management, and optimization of the power system [15–22], etc. in the smart grid and have briefly described their impacts on reliability. This chapter reviews load management in the future smart grids and its effect on power system reliability. It also discusses about the need to define new indices for reliability evaluation in the smart grid environment. This chapter is organized as follows. Section 2 describes the load management in the smart grids from two viewpoints: DR and LC. Section 3 explains direct and indirect effects of the load management on the smart grid reliability. A discussion about the necessity of defining new reliability evaluation indices in the smart grid environment is presented in Sect. 4. Also, some new reliability indices are proposed in this section. Conclusions are drawn in Sect. 5.

2 Load Management in Smart Grid

Load management can be divided into two separate forms of actions: DR and LC. In fact, DR is the customer-side of load management, and LC is the utility-side of load management. From the viewpoint of DR, load management means control and optimization of demand by customers to reduce their electricity bills and increase their income by participating in the power system management programs. On the other hand, from the viewpoint of LC, load management means control and optimization of demand by system operator (SO) to guarantee the power system security and reliability, optimize the asset utilization, and mitigate the contingencies in power system.

2.1 Demand Response in Smart Grid

According to the US Department of Energy (DOE) report in 2006, DR is "changes in electric usage by end-use customers from their normal consumption patterns in response to changes in the price of electricity over time, or to incentive payments designed to induce lower electricity use at times of high wholesale market prices or when system reliability is jeopardized" [23]. The most conventional model of load management is utility DLC, which has been implemented for a long time in some form or the other. It is obvious that such procedure is not acceptable and popular among customers and reduces the quality of service, which is an essential factor in power markets efficiency. Smart grid would employ digital technologies to provide two-way, real-time information and communications over the grid (Fig. 1). Such capabilities have a significant positive effect on DR programs. It is estimated that in 2019, existing DR programs could reduce the US peak demand by 4 %, while using dynamic pricing programs for all electric consumers (as a result of the emerging smart grids) will decrease peak demand by 20 % [9].

Electricity markets are based on the assumption that the electrical energy can be treated like a commodity and so it can be traded in competitive markets. Therefore, consumers in an electricity market are expected to raise their demand up to the point that there is a balance between the marginal value and the cost of electricity. It is obvious that the flat-rate tariffs for the electricity consumption could not motivate the customers to reply to the market prices. In fact, customers who pay as a flat-rate will be isolated from the spot prices of the electricity. However, the customers in the present electricity markets will not change their demand so much in response to short-term increases in the electricity prices, even if they face with the spot prices [11]. In other words, in the existing electricity markets the short run price elasticity of demand is small because of economical and social considerations [11]. The poor elasticity of the demand causes large price spikes in the electricity markets and can result in exerting market power by generating companies [11].

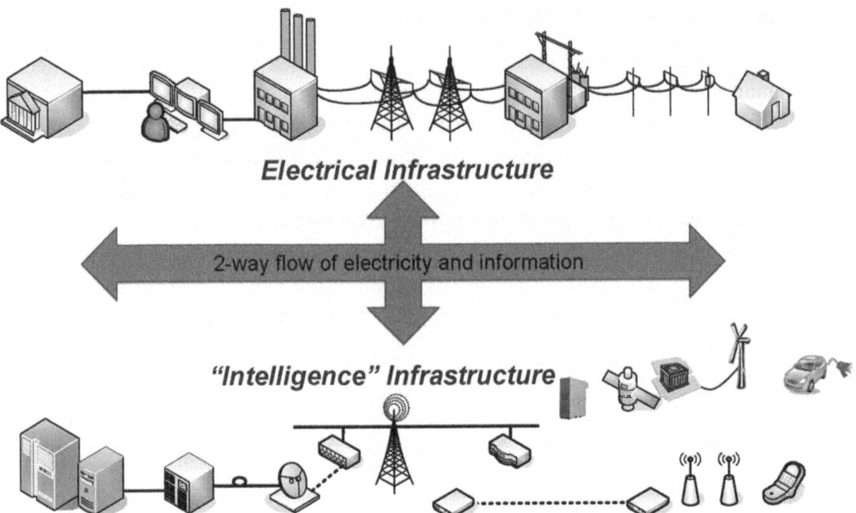

Fig. 1 Power and information flow under smart grid (Source: The NIST Interoperability Framework for Smart Grid)

A piecewise linear curve with three segments can be used to represent a typical supply function in an electricity market (Fig. 2) [24]. The first section depicts the bulk generating units in a competitive market. The peaking units, which are rarely committed to the grid, are represented by the second section of the supply curve with much steeper slope. The third section is vertical and shows the situation which all the existing generation capacity is in use. The demand function with low elasticity can be represented by an almost vertical line. This line moves horizontally as the demand varies over time. Two demand curves corresponding to a low demand scenario and a peak demand scenario have been showed in Fig. 2 [24]. The intersection of the supply function with each of these curves determines the electricity price for that scenario. As depicted in Fig. 2, the price rises sharply during periods of peak demand because the marginal producer will be a peaking unit. As shown in Fig. 3, in the cases that the whole reserve capacity is in use during peak load condition, the electricity price increases enormously and a price spike occurs. Therefore, due to this poor elasticity of demand, generating companies can exert market power by deliberately reducing the generation capacity.

As shown in Fig. 4, if a part of demand-side has an acceptable elasticity of demand and can properly reply to the price variations, the demand curve will consist of two parts [25]. The first section, which is depicted by a vertical line, is the price taking part and cannot respond to the electricity price variations. The second section with a negative slope shows the price responsive part of demand which represents DR. As can be seen in Fig. 4, the price responsive part of demand can mitigate price spikes by reducing the demand during critical conditions. The price taking part of demand will also benefit from this enabled DR which causes reduced price of electricity in the market.

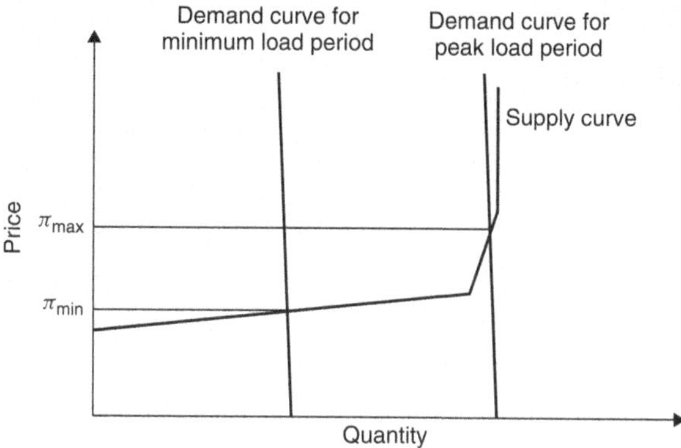

Fig. 2 Electricity price for low demand and peak demand scenarios [24]

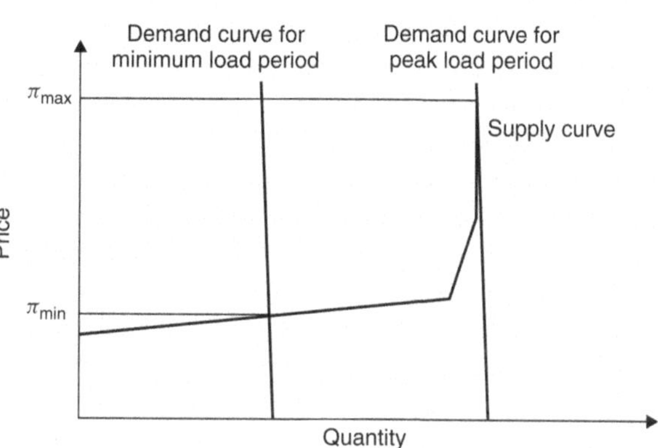

Fig. 3 The mechanism which leads to price spikes in the markets due to insufficient electricity generation [24]

The price responsive part of demand in the most of the existing electricity markets is very small. Usually, only large consumers can directly participate in the wholesale electricity market. In general, small consumers face two practical problems for direct participation in the competitive electricity markets. The first issue is the need for continuous measurement of real-time consumption of consumers. Also, it is necessary to notify the consumers about the electricity prices of the upcoming periods. In the existing power grids, the cost of implementing such an advanced metering and communication systems will be very high. The benefits which the small consumers gain by participating in a competitive electricity market may not recover these costs. The second problem is the number of small consumers in a power system. Since there are millions of consumers in a power system, the amount of data which should be sent and received by market operators will be unimaginable [11].

The role of consumers and DR in the electricity markets will be improved in the future smart grids. Advanced technologies like AMI, smart controllers, and smart monitors will have significant impacts on DR. AMI systems can provide the smart metering and real-time measurement of electricity consumption in the smart grid environment. Therefore, the spot prices of electricity can be applied to

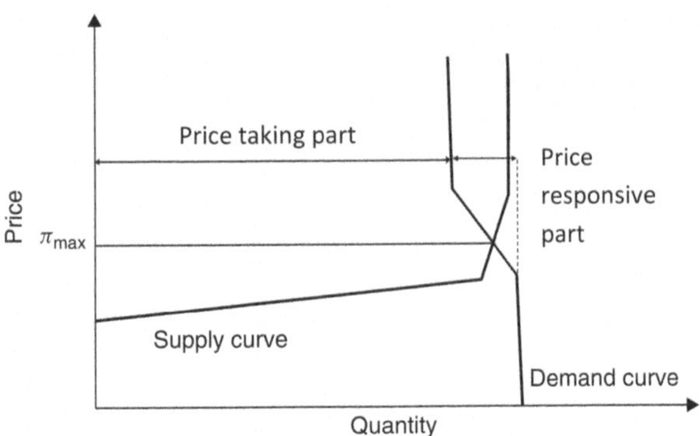

Fig. 4 The effect of DR on price spikes in the electricity markets [11]

all consumers based on the real-time consumption measurements. Facing the electricity price variations will encourage customers to manage their demand based on the market signals. Smart monitors are the other side of this mechanism which can provide the necessary information (such as the spot prices of electricity and the system operator messages) for consumers. With this gadget, the consumers can receive the details of their consumption and the price for upcoming periods and therefore will be encouraged to reduce or shift their demand during the periods of high prices. Therefore, AMI systems in conjunction with the smart monitors can enhance DR in the electricity markets.

In the smart grid environment, with the advent of smart controllers and smart sensors, demand can be efficiently managed in order to reduce the cost of electricity for consumers or to meet the power system security requirements for the SO. This mechanism can be employed in local or global modes. In the local mode, this smart managing system will be arranged based on the consumer's decisions and desires, while in the global mode, the managing system of each consumer will communicate with its higher level controllers in a hierarchical configuration. This global management system can be employed by a retailer to manage the financial risks; or by the SO to meet the power system security requirements [11]. Therefore, development of the smart grid's technologies will improve DR and can inspire small consumers for active participation in the future electricity market.

2.2 Load Control in Smart Grid

Direct control of large amounts of customer loads will provide an effective tool for addressing key smart grid issues like self-healing and optimized power flow. This ability will facilitate wide utilization of intermittent generation (wind and solar power generation) and dealing with increasing energy consumption due to emerging electric vehicles technology in the future. The future smart grid technologies like smart sensors, smart controllers, smart appliances, and AMI will effectively enable DLC.

Table 1 shows the share of kilowatt hour consumption for a typical feeder [8]. In this case, as much as almost 48 % of the load could be interrupted in the emergency conditions with no noticeable impact on consumers. In shorter periods, up to

Table 1 Share of kWh consumption for a typical feeder [8]

Appliance	Percentage (%)	Consumption Percentage (%)
Electric heat	39.9	39.9
Furnace fans	8.9	48.8
Electric water heat	7.5	56.3
Refrigerator	6.6	62.9
Freezer	1.2	64.1
Clothes washer	0.2	64.3
Clothes dryer	2.7	67.0
Dishwasher	1.6	68.6
Range	1.9	70.5
TV	3.1	73.6
Incandescent lighting	26.4	100
Total	100	–

68 % of the load could be interrupted with no significant influences on consumers' essential needs.

It is clear that the amounts of acceptable emergency load interruption using DLC vary from one area to another. Also, it depends on many factors such as geographical characteristics, electricity consumption culture, income, and prosperity in each region. However, DLC is a powerful tool for SO to control and mitigate contingencies and therefore to improve power system security and reliability. Thanks to the future smart appliances and AMI, DLC can be implemented as smart load control (SLC) which is so different from inefficient existing load shedding schemes. Conventional load shedding schemes curtail the entire load of each consumer, whereas in SLC only the low priority parts of each consumer's load will be interrupted. In this procedure, the priority of different load sections for each consumer can be determined by SO (or even by the consumer itself). Such method provides several technical benefits such as increasing the consumers' satisfaction, decreasing the value of lost load (VOLL), improving the flexibility of power system to deal with contingencies, and providing wider options for SO in the emergency conditions.

3 Load Management and Smart Grid Reliability

Renewable resources, load management, electric storage, and electric transportation are some of the most important resources of the smart grid which will have significant effects on smart grid reliability [3]. While renewable resources increase the generation capacity and decline the environmental problems in the power system, they aggravate reliability because of their volatility. Electric storages have economic benefits for the grid and can enhance reliability through reducing peak demand and load variability. Also, electric transportation is a solution for the environmental concerns and has the potential to reduce load variability. However, balancing these resources to maintain reliability will be challenging due to their diverse characteristics and it needs a sophisticated grid-wide implementation of communication and information technologies. Detailed investigation of the effects of these resources on smart grid reliability is out of the scope of this chapter. However, the impacts of the other resource, i.e., load management, on smart grid reliability will be reviewed in more detail here. The direct and indirect effects of load management have been discussed below.

3.1 Direct Effects of Load Management on Reliability

Load management can be defined as the reduction of load in the emergency conditions or when the electricity price is high. These conditions usually occur during peak load periods or congested operation. DR is usually referred to the load reduction which is initiated by the consumer. Nonemergency DR can reduce the need for additional resources and can decrease the real-time electricity prices. Generally, energy conservation is not the main goal of DR because a large part of the reduced energy during load curtailment will be consumed at a more appropriate time. Therefore, DR will usually result in a flatter load profile. This capability to flatten the load profile will be improved by increasing the participation of consumers in electricity market and DR programs. Due to the pre-mentioned problems, price-based DR has not been widely implemented in the power grids yet and the contract-based participation of consumers is also very low [3]. As mentioned before, by emerging smart grids, DR can be effectively implemented and customers can participate in the competitive electricity markets using advanced smart grid technologies. DR can also serve as an ancillary service to enhance reliability. Also, SLC is an efficient method to deal with emergency conditions and to improve system reliability with as less as possible impacts on consumers.

Under ideal conditions where DR, storage, and electric vehicles can be completely coordinated with all other resources, the load profile would be almost flat (Fig. 5). So, the grid would be operated near the peak condition most of the times. At first, this situation will result in improved system reliability because of the reduction in peak demand. However, as the flattened load grows over time, the system will be forced to operate closer to the edge due to optimal asset utilization considerations. Such condition puts the system at risk more often and makes it more vulnerable to failures [3]. In other words, in such condition, lost of a sufficient portion of generation will result in generation inadequacy for a large period of time, whereas, in the existing power systems which the load profile is not so flat, losing a portion of generation will probably result in generation inadequacy only in peak load periods.

3.2 Indirect Effects of Load Management on Reliability

SLC is an effective mechanism to deal with contingencies and emergency conditions which can improve power system stability. Also, by improving the stability of power system, the system security and reliability will be enhanced. Therefore, SLC, which can be enabled in the future smart grids, will indirectly improve the power system reliability.

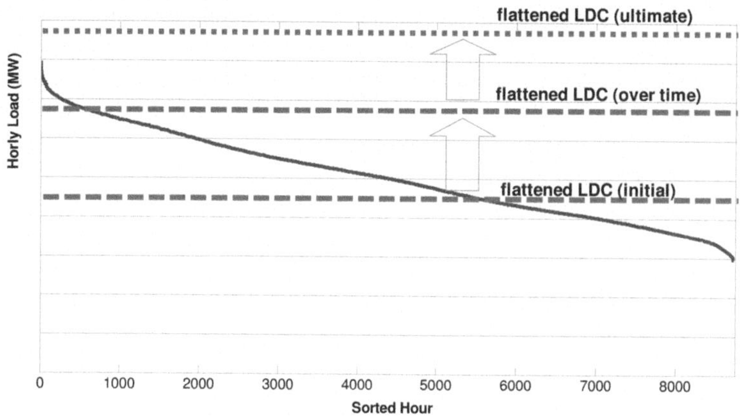

Fig. 5 Ultimate reliability impacts of smart grid resources [3]

Transient angle stability can be improved by direct control of local loads. This can be achieved by decreasing the load in areas where generators are decelerating, and increasing the load in areas where generators are accelerating. Loads can be simply decreased by opening their switches, but it is not easy to immediately increase local loads. It is mentioned in [8] that the load of heating and cooling systems can be temporarily increased by overriding their thermostats. For example, if the supply of a single phase refrigerator motor is interrupted for about 0.2 s, it will stall and then draw six times the normal power. Although in this example the equipment could be damaged, the point is, for identifying the potential of load components to enhance system reliability, a detailed study on their behavior will be necessary [8]. Frequency instability usually occurs when the system has separated into multiple islands. The frequency usually falls after a few seconds in the islands which the generation is insufficient. To stabilize the frequency and prevent generators from under-frequency tripping, fast load control should be implemented. On the other hand, the frequency will rise after a few seconds for islands with extra generation. Fast load controls can be also implemented in this case to increase the load and to prevent generators from over-frequency tripping [8].

Voltage collapse occurs in the form of transient, midterm, or long-term. A power swing or a fault which dips the voltage can cause transient voltage collapse. Transient voltage collapse reduces the speed of induction motors and leads to increase in their demands. Such induction motors should be rapidly disconnected from grid to prevent voltage instability [8]. This procedure can be implemented successfully using SLC. Midterm or classical voltage collapse usually occurs due to very high loads, high dependency on remote generation, or a sudden disturbance like generators or transmission lines outages. In this case, automatic load control methods that can curtail the load within a few seconds should be employed to prevent midterm or classical voltage collapse [8]. In the future smart grids, SLC will be a flexible method which can reduce load with as less as possible impacts on consumers. In the case of long-term voltage collapse, a very large load increase or large power transfers can initiate voltage instability. Poor support of reactive power from local generators causes long-term voltage collapse. Load shedding is one of the corrective actions which can be employed to prevent this type of voltage collapse [8]. However, enabling SLC by smart grids will provide a more flexible tool for load interruption compared with the conventional inefficient load shedding schemes.

According to [8], "Thermal overload stability refers to lines exceeding their thermal rating, sagging, flashing over, and then tripping." In this case, the lload shedding should be performed within a few seconds to alleviate thermal line and transformer overloads. Automatic load controls schemes like SLC can also be employed to prevent thermal overload stability. Therefore, future smart grids will provide more effective and more flexible tools and resources to meet the power system stability requirements.

4 Reliability Evaluation Indices in Smart Grid

It seems that some issues in reliability evaluation area should be reconsidered in the future smart grids environment. First of all, the load modeling in classic methods of reliability evaluation should be modified. Since in the smart grid environment, the load behaves completely dynamic and many factors like the electricity price affect it, static models for load will not be useful anymore. Moreover, as mentioned before, by emerging smart grid, the load duration curve will totally change and therefore, it will be necessary to develop new models for the future load duration curves. However, these issues are out of the scope of this chapter.

The next issue is the necessity of defining new indices for reliability evaluation due to significant changes in the LC methods. The following sections discuss the reliability indices of distribution system in the future smart grids.

4.1 Reliability Indices in Distribution System

Power system reliability indices are the performance measures of reliability which are employed by electric utility industry. The reliability indices mainly include measures of outage duration, frequency of outage, the amount of power or energy which is not supplied, and the number of customers involved in outages. The Institute of Electrical and Electronic Engineers (IEEE) defines the generally accepted reliability indices in its standard: IEEE Std. 1366. The interruption indices like system average interruption frequency index (SAIFI), system average interruption duration index (SAIDI), and customer average interruption duration index (CAIDI), and energy-based indices like energy not supplied (ENS), are the main distribution reliability evaluation indices which are used in power system studies.

"The SAIFI indicates how often the average customer experiences a sustained interruption over a predefined period of time, usually a year" [26]:

$$\mathrm{SAIFI} = \frac{\sum_i N_i}{N_T} \tag{1}$$

where the sum is taken over all interruption events i, N_i is the number of customers interrupted at event i, and N_T is total number of customers in the system.

"SAIDI indicates the total duration of interruption for the average customer during a predefined period of time" [26]:

$$\text{SAIDI} = \frac{\sum_i N_i \cdot r_i}{N_T} \tag{2}$$

where r_i is the restoration time for interruption event i.

"CAIDI represents the average time required to restore service" [26]:

$$\text{CAIDI} = \frac{\sum_i N_i \cdot r_i}{\sum_i N_i} = \frac{SAIDI}{SAIFI}. \tag{3}$$

"*ENS* gives the total amount of energy that would have been supplied to the interrupted customers if there would not have been any interruption" [26]:

$$\text{ENS} = \sum_i P_i \times r_i \tag{4}$$

where P_i is the average load interrupted by each interruption i.

4.2 The Need to Modify Reliability Indices

As mentioned before, emerging smart grids and development of their related advanced technologies will enable the DLC of all consumers over the grid. In the emergency conditions in which SO should interrupt a portion of electricity load, except in the cases that fault occurs on physical connections such as feeders, smart load control can be performed instead of conventional load shedding. In other words, load interruption will be limited to the low priority loads and the essential part of each consumer's load, like lighting, will be supplied continuously. This method will improve the quality of service, increase the consumers' satisfaction, and enhance system efficiency. Therefore, the first step is to modify reliability indices in order to involve the priority and the amount of interrupted parts of load for each customer.

The electric distribution system of Thomas Edison in late nineteenth century was based on direct current (DC) power generation, delivery, and use. However, today's distribution systems are mainly based on alternating current (AC) due to the significant advantages of AC power over DC power in transforming and transmission. But, there is a newfound interest in DC power delivery systems due to several potential benefits of such systems [2]:

- There is an increasing number of equipments which operate on DC power and need conversion from AC sources.
- The output of most of the distributed generation systems is in DC form.
- Batteries, flywheels, capacitors, and other storage devices store and deliver DC power.
- Electric transportation and PHEVs also use DC power.
- Energy efficiency in data centers can be enhanced by DC power.
- Advanced inverters and power electronics have facilitated efficient conversion of DC power to AC power of different voltage levels.
- The integration, operation, and performance of microgrid systems can be enhanced by DC power delivery.

Therefore, it seems that in the future smart grids, delivering of the electricity will be in both AC and DC forms. Since the main purpose of reliability indices is to identify the weak points of the network, which will be a combination of AC and DC systems, it will be better to differentiate between AC and DC indices. As a result, the reliability indices shown in (1) to (4) need some modifications to make them compatible with the future smart grids:

$$\mathrm{SAIFI}^{\mathrm{AC/DC}} = \frac{\sum_i \sum_j \sum_k n_{j,k}^{AC/DC} \cdot p_{i,j,k}^{AC/DC}}{\sum_j \sum_k n_{j,k}^{AC/DC} \cdot p_{j,k}^{AC/DC}} \tag{5}$$

where the sum is taken over all events i, all consumers j, and all priorities k. $n_{j,k}^{AC/DC}$ is the penalty factor for interrupting a portion of j_{th} consumer's AC/DC load which has the priority of k. $p_{i,j,k}^{AC/DC}$ is the j_{th} consumer's AC/DC load part with priority of k which has been interrupted in i_{th} interruption. $p_{j,k}^{AC/DC}$ is the j_{th} consumer's AC/DC load part with priority of k. Note that $p_{i,j,k}^{AC/DC}$ and $p_{j,k}^{AC/DC}$ are fractions and they range are from 0 (nothing of the whole consumer's load) to 1 (the whole consumer's load).

$$\mathrm{SAIDI}^{\mathrm{AC/DC}} = \frac{\sum_i \sum_j \sum_k n_{j,k}^{AC/DC} \cdot p_{i,j,k}^{AC/DC} \cdot r_{i,j,k}^{AC/DC}}{\sum_j \sum_k n_{j,k}^{AC/DC} \cdot p_{j,k}^{AC/DC}} \tag{6}$$

where $r_{i,j,k}^{AC/DC}$ is the restoration time for the j_{th} consumer's AC/DC load part with priority of k at i_{th} interruption.

$$\mathrm{CAIDI}^{\mathrm{AC/DC}} = \frac{\sum_i \sum_j \sum_k n_{j,k}^{AC/DC} \cdot p_{i,j,k}^{AC/DC} \cdot r_{i,j,k}^{AC/DC}}{\sum_i \sum_j \sum_k n_{j,k}^{AC/DC} \cdot p_{i,j,k}^{AC/DC}} \tag{7}$$

$$\mathrm{ENS}^{\mathrm{AC/DC}} = \sum_i \sum_j \sum_k n_{j,k}^{AC/DC} \cdot P_{i,j,k}^{AC/DC} \cdot r_{i,j,k}^{AC/DC} \tag{8}$$

where $P_{i,j,k}^{AC/DC}$ is the j_{th} consumer's AC/DC average load portion with part of k which has been interrupted in i_{th} interruption. Note that $P_{i,j,k}^{AC/DC}$ is the real value of the interrupted load in W/ kW.

5 Conclusion

This chapter reviewed the effects of advanced load management on smart grid reliability. Load management is divided into DR and LC activities. As discussed in this chapter, emerging smart grid and development of advanced technologies such as AMI, smart monitors, smart appliances, and smart controllers, along with an advanced communication infrastructure can facilitate active participation of small consumers in the electricity market and will improve DR. In the future smart grids, under ideal conditions, DR, storage, and electric vehicles will be closely coordinated with all other resources such that the net load profile would be nearly flat and this can make the system more Vulnerable to failure.

Also, the ability of direct load control of millions of customers in the power networks, which will be enabled by emerging smart grids, will facilitate the transition from conventional inefficient load shedding schemes to the efficient smart load control methods. In SLC, only the low priority parts of demand will be interrupted in emergency conditions, and the essential needs of consumers will be continuously supplied. SLC is a powerful method to deal with contingencies and emergency conditions which can improve power system stability and therefore improve smart grid security and reliability. Due to significant changes in load control methods, it is necessary to modify existing reliability indices to be applicable in the future smart grids. In this chapter, some changes in the reliability indices of distribution systems were discussed and some modifications were proposed.

References

1. Ipakchi A, Albuyeh F (2009) Grid of the future. IEEE Power Energy Mag 7(2):52–62
2. Gellings CW (2009) The smart grid: enabling energy efficiency and demand response. CRC Press
3. Moslehi K, Kumar R (2010) A reliability perspective of the smart grid. IEEE Trans Smart Grid 1(1):57–64
4. Santacana E, Rackliffe G, Tang L, Feng X (2010) Getting smart. IEEE Power Energy Mag 8(2):41–48
5. Collier SE (2010) Ten steps to a smarter grid. IEEE Ind Appl Mag 16(2):62–68
6. Amin SM, Wollenberg BF (2005) Toward a smart grid: power delivery for the 21st century. IEEE Power Energy Mag 3(5):34–41
7. Lightner EM, Widergren SE (2010) An orderly transition to a transformed electricity system. IEEE Trans Smart Grid 1(1):3–10
8. Podmore R, Robinson MR (2010) The role of simulators for smart grid development. IEEE Trans Smart Grid 1(2):205–212

9. Hamilton K, Gulhar N (2010) Taking demand response to the next level. IEEE Power Energy Mag 8(3):60–65
10. Rahimi F, Ipakchi A (2010) Demand response as a market resource under the smart grid paradigm. IEEE Trans Smart Grid 1(1):82–88
11. Moshari A, Yousefi GR, Ebrahimi A, Haghbin S (2010) Demand-side behavior in the smart grid environment In: Proceedings. IEEE PES conference on innovative smart grid technologies Europe, Sweden, Oct 2010
12. Liserre M, Sauter T, Hung JY (2010) Future energy systems- integrating renewable energy sources into the smart power grid through industrial electronics. IEEE Ind Electron Mag 4(1):18–37
13. Su SY, Lu CN, Chang RF, Alcaraz GG (2011) Distributed generation interconnection planning: a wind power case study. IEEE Trans Smart Grid 2(1):181–189
14. Thomas RJ (2009) Putting an action plan in place. IEEE Power Energy Mag 7(4):26–31
15. Russell BD, Benner CL (2010) Intelligent systems for improved reliability and failure diagnosis in distribution systems. IEEE Trans Smart Grid 1(1):48–56
16. Hedman KW, Ferris MC, O'Neill RP, Fisher EB, Oren SS (2010) Co-optimization of generation unit commitment and transmission switching with N-1 reliability. IEEE Trans Power Syst 25(2):1052–1063
17. Bouhouras AS, Andreou GT, Labridis DP, Bakirtzis AG (2010) Selective automation upgrade in distribution networks towards a smarter grid. IEEE Trans Smart Grid 1(3):278–285
18. Meliopoulos APS, Cokkinides G, Huang R, Farantatos E, Choi S, Lee Y, Yu X (2011) Smart grid technologies for autonomous operation and control. IEEE Trans Smart Grid 2(1):1–10
19. Desai B, Lebow M (2010) Needed: asap approach. IEEE Power Energy Mag 8(6):53–60
20. Varaiya PP, Wu FF, Bialek JW (2011) Smart operation of smart grid: risk-limiting dispatch. Proc IEEE 99(1):40–57
21. Zhang P, Li F, Bhatt N (2010) Next-generation monitoring, analysis, and control for the future smart control center. IEEE Trans Smart Grid 1(2):186–192
22. Heydt GT (2010) The next generation of power distribution systems. IEEE Trans Smart Grid 1(3):225–235
23. Mansueti L (2006) DOE's EPACT report to congress on demand response in electricity markets, US Dept of Energy rep, 13 Mar 2006
24. Kirschen D, Strbac G (2004) Fundamentals of power system economics. Wiley, New York
25. Su CL, Kirschen D (2009) Quantifying the effect of demand response on electricity markets. IEEE Trans Power Syst 24:1199–1207
26. Cepin M (2011) Assessment of power system reliability. Springer

Evaluating the Performance of Small Autonomous Power Systems Using Reliability Worth Analysis

Marios N. Moschakis, Yiannis A. Katsigiannis and Pavlos S. Georgilakis

1 Introduction

A small autonomous power system (SAPS) is a system that generates electricity in order to serve a nearby low energy demand, and it usually operates in areas that are far from the grid. Generally, there are three methods of supplying energy in rural areas: grid extension, use of fossil fuel generators, and hybrid power systems with renewable energy sources (RES). In isolated or remote areas, the first two methods can be very expensive [1]. The typical cost of a low-voltage distribution line is about US$ 3,000/km for the plains and it increases by 10–25 % for remote hilly regions [2], whereas the cost of fossil fuel delivery in these areas may be greater than the cost of the fuel itself.

RES can often be used as a primary source of energy in such a system, as they are usually present in geographically remote and demographically sparse areas. However, since renewable technologies such as wind turbines (WTs) and photovoltaics (PVs) are dependent on a resource that is not dispatchable, there is an impact on the reliability of the electric energy of the system, which has to be considered [3]. The basic way to solve this problem is to use storage and/or dispatchable generators, such as diesel generators.

M. N. Moschakis (✉)
Department of Electrical Engineering, Technological Educational Institute of Larissa, Larissa, Greece
e-mail: mmoschakis@teilar.gr

Y. A. Katsigiannis
Department of Natural Resources & Environment, Technological Educational Institute of Crete, Chania, Crete, Greece
e-mail: katsigiannis@chania.teicrete.gr

P. S. Georgilakis
School of Electrical and Computer Engineering, National Technical University of Athens (NTUA), Athens, Greece
e-mail: pgeorg@power.ece.ntua.gr

R. Karki et al. (eds.), *Reliability Modeling and Analysis of Smart Power Systems,* Reliable and Sustainable Electric Power and Energy Systems Management, DOI 10.1007/978-81-322-1798-5_10, © Springer India 2014

Due to the unique characteristics of SAPS, reliability evaluation is crucial in these systems [1, 4]. The most traditional methods for the reliability evaluation of SAPS are mainly deterministic techniques. However, these techniques do not define consistently the true risk of the system, as they can lead to very divergent risks even for systems that are very similar [5]. In addition, these techniques cannot be extended to include intermittent sources, such as wind energy [6]. A second approach for reliability evaluation of power systems is direct analytical methods. These methods overcome the problems of deterministic techniques, but they cannot completely recognize the chronological variation of intermittent sources, such as wind and solar energy. These factors can be incorporated using the Monte Carlo simulation (MCS), which however increases significantly the computation time.

This chapter investigates the effect of reliability worth on the optimal economic operation of SAPS that is based on RES technologies. The location of the studied system is in Chania region, Greece. The optimization procedure is implemented with a combined genetic algorithm (GA) and local search procedure. GA is a powerful optimization technique that has been proposed for the solution of a variety of problems, including optimal SAPS sizing [7–9] and distributed generator placement in power distribution networks [10]. In the optimization procedure, the objective function is the minimization of SAPS cost of energy (in €/kWh), and three scenarios are examined: (i) no consideration of reliability worth, (ii) consideration of reliability worth for agricultural load type, and (iii) consideration of reliability worth for residential load type. In addition, this chapter examines the effect of considering SAPS components forced outage rate in the obtained optimal solutions for the above three examined scenarios. This analysis, which is implemented via MCS, aims to highlight the difference between the results obtained from a typical SAPS optimization procedure (e.g., [7–9, 11]), and the results of an approach that takes into account reliability issues related to the operation of the studied system. This procedure is repeated for a large number of alternative scenarios, in order to study the effects for a large number of key and uncertain parameters.

The chapter is organized as follows. Section 2 presents information about reliability analysis of power systems, as well as details about the calculation of reliability worth. Section 3 formulates the optimization problem, whereas Sect. 4 presents SAPS modeling details. Section 5 provides a brief description of the examined system and compares the results of the optimization procedure and the MCS. Section 6 presents the results of sensitivity analysis and Sect. 7 concludes the chapter.

2 SAPS Reliability Analysis

A variety of probabilistic indices can be calculated, in order to evaluate the performance of a power system in a reliability framework. The two basic probabilistic indices used are the loss of load expectation (LOLE) and the loss of energy expectation (LOEE) [5]. LOLE is defined as the average number of hours for which the

Table 1 CDF values (€/kW)

User sector	Interruption duration			
	20 min	1 h	4 h	8 h
Agricultural	0.2541	0.4807	1.5289	3.0519
Residential	0.0689	0.3570	3.6400	11.6222

load is expected to exceed the available capacity. On an annual basis, LOLE can be expressed mathematically as:

$$\text{LOLE} = \Delta t \bullet \sum_{\Delta t} t_{outage}(i) \tag{1}$$

where $t_{outage}(i)$ is equal to 1 for the case that the load in time step i is greater than the generating capacity plus the battery storage level and 0 otherwise. LOEE is defined as the expected energy (in kWh) that will not be supplied when the load exceeds the available generation, and can be expressed as:

$$\text{LOEE} = \Delta t \bullet \sum_{\Delta t} e_{unserved}(i) \tag{2}$$

where $e_{unserved}(i)$ is the energy not supplied in the time step i of the year. However, the actual benefits of a power system's operation can only be assessed by conducting relevant cost and reliability studies. It is therefore important to determine the optimal reliability level at which the reliability investment achieves the best results in reducing the customer damage costs due to power supply interruptions. This approach can be expressed mathematically as the minimization of total cost, which is equal to the sum of life cycle cost and customer damage cost.

For the calculation of the expected customer damage cost, the customer damage functions (CDFs) have been used. The CDF is an index (expressed mainly in $/kW) that depends on the type of user and the interruption duration. There are a few studies that contain interruption cost data. Reference [4] contains data for the power utilities of Canada. Similar studies in Greece [12] have shown coincidence with the Canadian results. The values of CDFs, limited for the type of users that are considered in our study, are presented in Table 1. Interruption costs for durations different than the values shown in Table 1 were estimated using the same slope of the straight line joining the two nearest duration values of Table 1.

The CDF values can be converted into an extended index that links system reliability with customer interruption costs. One suitable form is the interrupted energy assessment rate (IEAR), expressed in €/kWh of unsupplied energy. The estimation of the IEAR indicates the severity, frequency and generation of the expected states of the generation model. In order to compute the IEAR, the expected customer interruption cost (ECOST) in €/year must be estimated first, taking into account the duration of interruption, the value of CDF and the unserved energy of each interruption. Then, IEAR can be calculated as follows:

$$\text{IEAR} = \frac{\text{ECOST}}{\text{LOEE}}. \tag{3}$$

For the investigation of SAPS performance, six reliability indices have been selected:

- LOLE.
- LOEE.
- Energy index of unreliability (EIU) that normalizes LOEE by dividing it with the annual energy demand.
- Frequency of interruptions (FOI), i.e., the expected number of times that loss of load occurs per year.
- Duration of interruptions (int), DOI, which is equal to LOLE/FOI, expressed in h/int.
- Energy not supplied index (ENSI) that is equal to LOEE/FOI, expressed in kWh/int.

3 Problem Formulation

The SAPS optimal sizing problem has to fulfill the objective defined by (Eq. 4) subject to the constraints (Eq. 6)–(Eq. 9). This problem is solved for three different scenarios: (i) no consideration of reliability worth, (ii) consideration of reliability worth for agricultural load type, and (iii) consideration of reliability worth for residential load type.

3.1 Objective Function

Minimization of system's cost of energy, $\min(COE)$.:

$$\min(COE). \tag{4}$$

The COE (€/kWh) of SAPS is calculated as follows:

$$COE = \frac{C_{antot}}{E_{anloadserved}} \tag{5}$$

where C_{antot} (€) is the total annualized cost and $E_{anloadserved}$ (kWh) is the total annual useful electric energy production. C_{antot} takes into account the annualized capital costs, the annualized replacement costs, the annual operation and maintenance (O&M) costs, and the annual fuel costs (if applicable) of system's components [11]. In case of considering customer damage costs, the value of COE includes IEAR.

3.2 Constraints

- Unmet load constraint [11]:

$$f_{UL} = \frac{\sum_{\Delta t}^{year} UL_{\Delta t} \bullet \Delta t}{E_{anload}} \leq f_{UL\max} \tag{6}$$

where f_{UL} is the annual unmet load fraction, $UL_{\Delta t}$ (kW) is the unmet load during the simulation time step Δt (h), E_{anload} (kWh) is the total annual electric energy demand, and $f_{UL\max}$ is the maximum allowable annual unmet load fraction. By its definition, f_{UL} is identical with EIU. In this chapter, the value of $f_{UL\max}$ has been taken equal to 5 %.
- Minimum renewable fraction constraint:

$$f_{RES} = \frac{E_{anRES}}{E_{antot}} \geq f_{RES\min} \text{ where } 0 \leq f_{RES\min} \leq 1 \tag{7}$$

where f_{RES} is the RES fraction of the system, E_{anRES} (kWh) is the total annual renewable energy production, E_{antot} (kWh) is the total annual energy production of the system, and $f_{RES\min}$ is the minimum allowable RES fraction. In this chapter, the value of $f_{RES\min}$ has been taken equal to 80 %. As a result, the energy production of studied SAPS is based mainly on RES technologies.
- Components' size constraints:

$$size_{comp} \geq 0 \quad \forall \ comp \tag{8}$$

$$size_{comp} \leq size_{comp\max} \quad \forall \ comp \tag{9}$$

where $size_{comp}$ is the size of system's component $comp$, and $size_{comp\max}$ is the maximum allowable size of $comp$. The values of $size_{comp\max}$ are shown in Table 2.

4 SAPS Components and Modeling

The considered SAPS has to serve electrical load, and it can contain the following component types:

- WTs.
- Polycrystalline silicon (poly-Si) PVs.
- Generator with diesel fuel.
- Lead-acid batteries.
- Converter.

Table 2 Component characteristics

Component	$size_{compmax}$	Increment	Capital cost	Replacement cost	O&M cost	Fuel cost	Lifetime
WTs (20 kW rated)	7 WT	1 WT	50,000 €/WT	40,000 €/WT	1,000 €/year	–	20 years
PVs	50 kW_p	1 kW_p	2,500 €/kW_p	2,000 €/kW_p	0	–	25 years
Diesel generator	50 kW	Variable	300 €/kW	300 €/kW	0.01 €/h per kW	1.5 €/L (diesel)	20,000 oper. hours
Batteries (1500 Ah,4V)	300 bat.	12 bat.	1,000 €/bat.	1,000 €/bat.	10 €/bat.	–	10,000 kWh
Converter	50 kW	1 kW	1,000 €/kW	1,000 €/kW	0	–	15 years

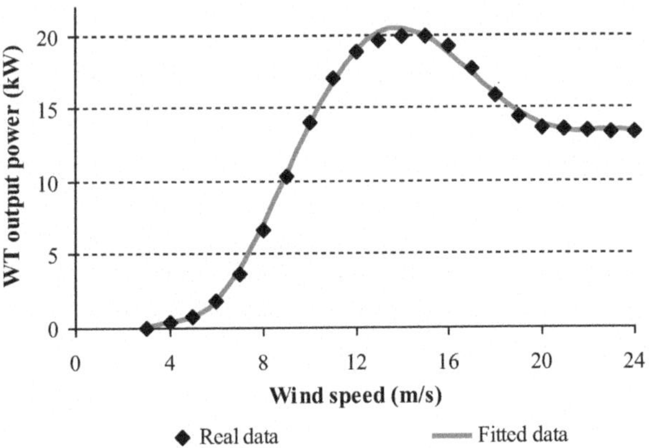

Fig. 1 Correlation between real and fitted data of WT power curve

The modeling of SAPS components is implemented as follows. The WT modeling is implemented using a power curve profile that is based on manufacturer's data. The selected WT has the following characteristics: rated power 20 kW AC, cut-in speed (V_{in}) 3 m/s, and cut-out speed (V_{out}) 24 m/s. For the WT power curve fitting, a seventh order polynomial expression has been selected, as it provides accurate correlation with real data, while it presents exclusively positive values for the generated power in the interval $[V_{in}\ V_{out}]$. The correlation between power curve's real and fitted data is shown in Fig. 1.

The WT power curve refers to standard conditions at sea level, corresponding to a temperature of 15 °C (288.15°K) and air pressure of 101.325 kPa, resulting in a standard sea density ρ_{air0}=1.225 kg/m³ [13]. If the pressure and temperature conditions at the area of WT installation are different from those corresponding to the standard conditions, the resulting power from the WT power curve needs to be adjusted, multiplied by the following density ratio [14]:

$$\frac{\rho_{air}}{\rho_{air0}} = \left(\frac{Pr}{101.325}\right) \cdot \left(\frac{288.15}{273.15 + T}\right) \tag{10}$$

where ρ_{air} is the air density of the site (in kg/m^3), Pr is the air pressure of the site (in kPa), and T is the air temperature of the site (in °C). Air pressure decreases with elevation above sea level, and for an altitude up to 5,000 m can be approximated by [13]:

$$Pr = 101.29 - 0.011837 \cdot z + 4.793 \cdot 10^{-7} \cdot z^2 \tag{11}$$

where z is the altitude (in m).

In the PV modeling, the output of the PV array P_{PV} (in kW) is calculated from [15]:

$$P_{PV} = f_{PV} \cdot P_{STC} \cdot \frac{G_A}{G_{STC}} \cdot \left(1 + \left(T_C - T_{STC}\right) \cdot C_T\right) \tag{12}$$

where f_{PV} is the PV derating factor, P_{STC} is the nominal PV array power in kW$_p$ under standard test conditions (STC), G_A is the global solar radiation incident on the PV array in kW/m^2, G_{STC} is the solar radiation under STC (1 kW/m^2), T_C is the temperature of the PV cells, T_{STC} is the STC temperature (25 °C), and C_T is the PV temperature coefficient (−0.004/°C for poly-Si). The PV derating factor is a scaling factor applied to the PV array output to account for losses, such as dust cover, aging and unreliability of the PV array, and is considered to be equal to 0.80. T_C can be estimated from the ambient temperature T_a (in °C) and the global solar radiation on a horizontal plane G (in kW/m^2) using (Eq. 13) [16]:

$$T_C = T_a + \frac{(NOCT - 20)}{0.8} \cdot G \tag{13}$$

where $NOCT$ is the normal operating cell temperature, which is considered equal to 45 °C.

The diesel generator fuel consumption F (L/kWh) is assumed to be a linear function of its electrical power output [17]:

$$F = 0.08415 \cdot P_{rated} + 0.246 \cdot P \tag{14}$$

where P_{rated} is generator's rated power and P is generator's output power. Lead-acid batteries have been modeled assuming: (i) overall efficiency of 80 %, (ii) nominal voltage of 4V, (iii) nominal capacity (per unit) of 1,500 Ah (6 kWh), (iv) lifetime of 10,000 kWh, (v) minimum state of charge equal to 20 % of their nominal capacity, and (vi) maximum charge and discharge current equal to C/5. Finally, converter efficiency has been taken equal to 90 %.

The simulation process examines a particular system configuration, in which components sizes satisfy constraints (Eq. 8) and (Eq. 9). The necessary inputs for the simulation are: (i) annual time series data for wind speed, solar radiation, ambient temperature and load, (ii) component characteristics, (iii) constraint bounds, and (iv) general parameters (project lifetime, interest rate). The specific values for these data are described in Sect. 5.1. In the simulation, for every time step Δt, the available renewable power (from WTs and PVs) is calculated and then is compared with the load. In case of excess, the surplus renewable energy is charging the batteries, if they are not fully charged. If renewable power sources are not capable to fully serve the load, the remaining electric load has to be supplied by controllable generators and/or batteries. From all possible combinations, it is selected the one that supplies the load at the least cost. When the whole year's simulation has been completed, it is determined whether the system is feasible, i.e., it is checked if it satisfies the constraints (Eq. 6) and (Eq. 7). After the end of simulation, *COE* is calculated by taking into account: (i) the annual results of the simulation, (ii) the capital, replacement, O&M and fuel cost (if applicable) of each component, (iii) the ECOST (if considering CDFs), (iv) the components' lifetime, (v) the project lifetime, and (vi) the discount rate.

An additional aspect of system operation arises, which is whether (and how) the diesel generator should charge the battery bank. Two common control strategies that can be used are load following (LF) strategy and cycle charging (CC) strategy. It has been found [18] that over a wide range of conditions, the better of these two strategies is virtually as cost-effective as an ideal predictive strategy, which assumes the existence of perfect knowledge in future load and wind conditions. In the LF strategy, batteries are not charged at all with diesel-generated energy; the diesel operating point is set to match the instantaneous required load. LF strategy tends to be optimal in systems with a lot of renewable power, when the renewable power output sometimes exceeds the load. In the CC strategy, whenever the diesel generator needs to operate to serve the primary load, it operates at full output power. A setpoint state of charge, SOC_a, has also to be set in this strategy. The charging of the battery by the diesel generator will not stop until it reaches the specified SOC_a. In this chapter, three alternative values of SOC_a have been considered: 80 %, 90 % and 100 %, so the total number of examined dispatch strategies is four. CC strategy tends to be optimal in systems with little or no renewable power.

5 Results and Discussions

5.1 Case Study System

In the considered SAPS, the project lifetime and the discount rate are assumed to be 25 years and 5 %, respectively. The simulation time step is taken equal to 10 min (1/6 h). The annual wind, solar and ambient temperature data needed for the estimation of WT and PV performance refer to measurements for the mountainous region of Keramia (altitude 500 m), in Chania, Crete, Greece. The annual SAPS peak load

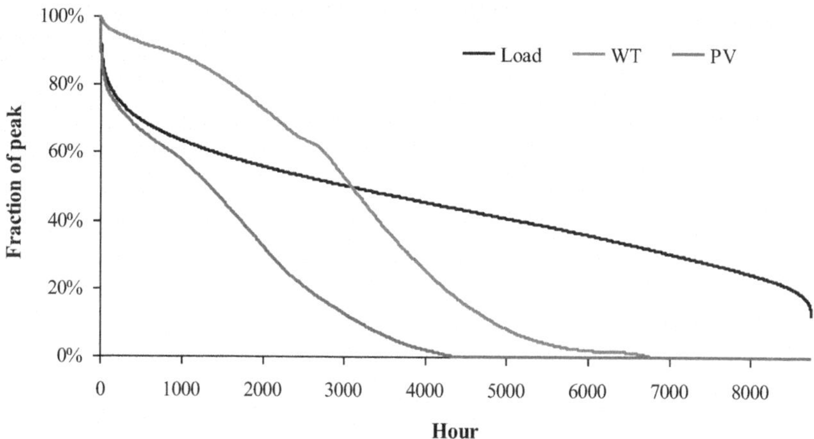

Fig. 2 Load, WT production, and PV production duration curves

has been considered equal to 50 kW, whereas the necessary SAPS load profile was computed by downscaling the actual annual load profile of Crete Island, which is the largest autonomous power system of Greece, with 600 MW peak load and 17 % min/max annual load. An additional noise has been added in the load profile, in order to reduce the min/max annual load ratio from 17 % (Crete power system) to 12 % (SAPS).

The considered values for anemometer height and WT hub height are 10 m and 35 m, respectively, assuming that power law exponent is equal to 0.20. Regarding PVs, it is considered that they do not include tracking system. The duration curves for load, WT production and PV production are depicted in Fig. 2.

The cost, lifetime, and size characteristics for each component are presented in Table 2. For each component, the minimum size is equal to zero. Moreover, with the exception of diesel generator, all components have constant increment of their size, as Table 2 shows. The considered sizes for the diesel generator are 0, 5, 10, 15, 25, 30, 40, and 50 kW. For the SAPS sizing problem of Table 2, the complete enumeration method requires:

$$\underbrace{8}_{\text{WTs}} \cdot \underbrace{51}_{\text{PVs}} \cdot \underbrace{8}_{\text{Dsl}} \cdot \underbrace{26}_{\text{Bat.}} \cdot \underbrace{51}_{\text{Conv.}} \cdot \underbrace{4}_{\text{Disp.}} = 17{,}312{,}256 \qquad (15)$$

i.e., over 17 million evaluations in order to find the optimal *COE*; in (Eq. 15) Disp. denotes the number of dispatch strategies. The computational time for each *COE* evaluation is 2.1 s. Consequently, the evaluations of the complete enumeration method require more than one year, for each one of the three considered scenarios. That is why it is essential to develop an alternative optimization method in order to solve the SAPS sizing problem in a fast and effective way.

5.2 GA Implementation for SAPS Optimal Sizing

Genetic algorithms (GAs) mimic natural evolutionary principles and constitute powerful search and optimization procedures. More specifically, binary GAs borrow their working principle directly from natural genetics, as the variables are represented by bits of zeros and ones. Binary GAs are preferred when the problem consists of discrete variables. The considered sizes of each SAPS component can take only discrete values, so the binary GA is proposed for the solution of SAPS optimal sizing problem.

In the binary GA, two alternative GA coding schemes can be used: conventional binary coding and Gray coding. In the proposed GA, each chromosome consists of six genes, of which the first five genes represent the SAPS component sizes (WT, PV, diesel generator, batteries, and converters), while the sixth gene refers to the adopted dispatch strategy (LF or CC). For the handling of constraints, the penalty function approach is adopted, in which an exterior penalty term is used that penalizes infeasible solutions. Since different constraints may take different orders of magnitude, prior to the calculation of the overall penalty function, all constraints are normalized.

The optimum configuration parameters of the adopted GA are: population size $N_{pop}=50$, number of generations $gn=15$, Gray coding, tournament selection, uniform crossover, and 0.01 mutation rate [8]. Additionally, the proposed GA is combined with local search procedure, in order to ensure that the selected solution is optimal compared to its neighbor solutions. Table 3 presents the optimal configurations and the six reliability indices for the three examined scenarios. As it can be seen, the consideration of no customer damage cost leads to a solution that presents the lowest *COE*. On the other hand, in this case the operation of SAPS is not the most reliable, since all reliability indices have their highest possible values in order the SAPS operation to be feasible, according to the problem constraints. The consideration of CDF increases the *COE* and improves significantly the reliability of the system by decreasing the PV size and increasing the diesel generator size. It can be seen that the consideration of either agricultural CDF or residential CDF provides almost identical results. This can be explained by the fact that agricultural CDF values are larger for small interruptions, but significantly lower for larger interruptions (more than 1 hour), as Table 1 shows. The optimal state is a compromise between these two situations, as reliability indices of Table 3 show. In all cases, the adopted dispatch strategy is LF, due to the large portion of RES technologies in energy production. The total number of performed objective function (*COE*) evaluations for the combined GA and local search procedure was 930 for all scenarios. Figure 3 shows the GA convergence for the three examined scenarios of Table 3.

Table 3 Optimal solutions of GA combined with local search

Scenario	WTs	PVs (kW_p)	Dsl(kW)	Batteries	Converter (kW)	Dispatch strategy	COE (€/kWh)
No customer damage cost	3	35	10	108	35	LF	0.2214
Agricultural CDF	3	50	40	144	40	LF	0.2659
Residential CDF	3	50	30	120	39	LF	0.2635
Scenario	LOLE (h/year)	LOEE (kWh/year)	EIU	FOI (int./year)	DOI (h/int.)	ENSI (kWh/int.)	
No customer damage cost	1053	9708.73	4.987 %	689	1.529	14.091	
Agricultural CDF	2.50	6.22	0.003 %	13	0.192	0.478	
Residential CDF	55.50	224.92	0.116 %	148	0.375	1.520	

5.3 Consideration of Components Forced Outage Rate

In the analysis of Sect. 5.2, no forced outage rate for any component of the system has been taken into account, in order to focus on the interruptions driven by the incapability of the system to meet the load demand. However, in order to evaluate more realistically the performance of the system, an analysis of components forced outage rate has to be included. This task is crucial especially for a SAPS, because there is no other way to supply its load other than by itself. The analysis is applied to the three optimal solutions shown in Table 3. For each one of them, a sequential MCS [19] is applied for a total number of 500 runs.

The consideration of forced outage rate is applied to the 2 SAPS components that contain rotating parts: WTs and diesel generator. For the WTs, a forced outage rate of 4 % for each WT has been considered, with mean time to failure (MTTF) equal to 1,920 h and mean time to repair (MTTR) equal to 80 h [6]. For the diesel generator, it is assumed that it needs scheduled maintenance every 1,000 h of operation. The duration of the maintenance follows uniform distribution in the hour interval [2 24]. Moreover, a starting failure of 1 % is included in the evaluation, while the repairing process follows the same distribution with the maintenance process [3].

The obtained results of MCS for the three examined cases are shown in Tables 4–6. These results include the minimum, maximum and average values, as well as the standard deviation of the six reliability indices and *COE*. Moreover,

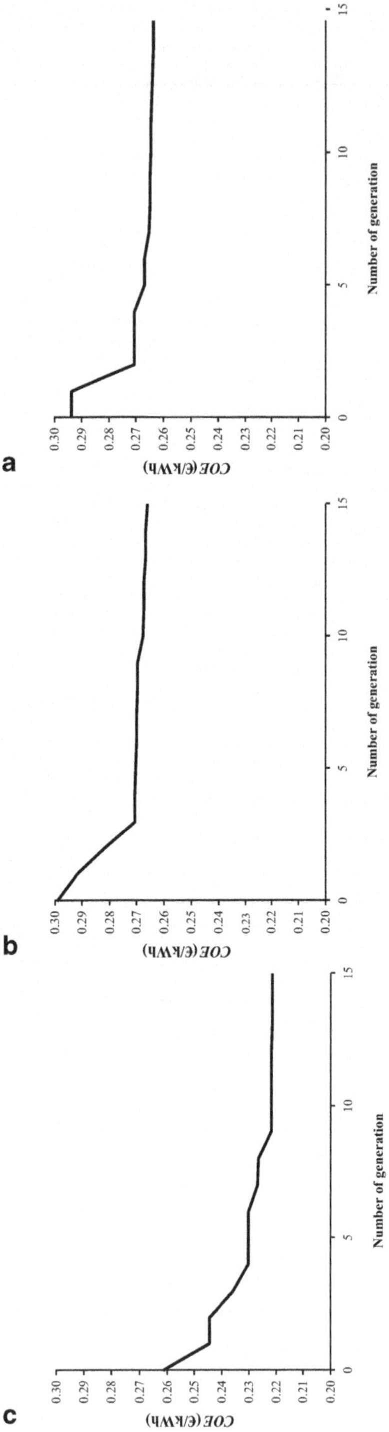

Fig. 3 GA convergence considering: **a** residential CDF, **b** agricultural CDF, **c** no customer damage cost

Table 4 MCS results considering no customer damage cost

Index	Min	Max	Average	Standard deviation	Coeffcient of variation
COE (€/kWh)	0.2228	0.2366	0.2286	0.0023	0.0102
LOLE (h/year)	1102.33	1654.67	1303.61	92.83	0.0712
LOEE (kWh/year)	9757.55	16478.74	11850.02	990.62	0.0836
EIU	5.012 %	8.464 %	6.086 %	0.509 %	0.0836
FOI (int./year)	669	1078	798.80	63.21	0.0791
DOI (h/int.)	1.437	1.849	1.634	0.0599	0.0367
ENSI (kWh/int.)	12.127	18.290	14.850	0.7894	0.0532

the (dimensionless) coefficient of variation is calculated, which is the ratio of the standard deviation to the mean, as a measure of variability. As it can be seen, the consideration of forced outage rate increases significantly the values of the basic reliability indices (LOLE, LOEE, EIU) and *COE*. In some cases, the values of the remaining reliability indices may be smaller compared to these of Table 2, but this does not mean that the performance is better. For example, the low values of FOI are combined with the large values of DOI and ENSI, resulting in lower number of interruptions that have higher duration.

Another interesting conclusion, drawn from the results shown in Table 4–6, is the higher variability (expressed by the coefficient of variation) of the basic reliability indices (LOLE, LOEE, EIU) and *COE*, in the scenarios of considering customer damage costs. In these two scenarios (agricultural and residential), the highest difference in variability is presented in *COE*, which can be explained by the fact that the residential customer damage cost is increased exponentially with the increase of interruption duration (see Table 1), affecting concurrently *COE*. Figures 4 and 5 present the variation of *COE* for these two scenarios.

6 Sensitivity Analysis

The uncertainty in many SAPS variables over which the designer has no control makes essential the need for sensitivity analysis. The uncertain parameters may contain weather data, and/or cost data. In this section, six alternative scenarios have been developed and analyzed. These scenarios are based on the following modifications of the case study system of Sect. 5.1 (initial scenario):

- 10 % increase of wind speed. In this scenario, the annual energy production of the WTs is increased by 9.08 %.
- 10 % decrease of wind speed. In this scenario, the annual energy production of the WTs is decreased by 11.67 %.
- 5 % increase of solar radiation. In this scenario, the annual energy production of the PVs is increased by 5.09 %.
- 5 % decrease of solar radiation. In this scenario, the annual energy production of the PVs is decreased by 5.27 %.

Table 5 MCS results considering agricultural CDFs

Index	Min	Max	Average	Standard deviation	Coefficient of variation
COE (€/kWh)	0.2673	0.3174	0.2867	0.0091	0.0319
LOLE (h/year)	94.50	443.67	210.16	63.64	0.3028
LOEE (kWh/year)	109.09	5479.53	1820.96	919.68	0.5051
EIU	0.056 %	2.814 %	0.935 %	0.472 %	0.5051
FOI (int./year)	450	793	566.40	62.51	0.1104
DOI (h/int.)	0.202	0.602	0.364	0.0720	0.1977
ENSI (kWh/int.)	0.221	7.579	3.098	1.2635	0.4078

Table 6 MCS results considering residential CDFs

Index	Min	Max	Average	Standard deviation	Coeffcient of variation
COE (€/kWh)	0.2649	0.3606	0.2965	0.0170	0.0573
LOLE (h/year)	139.50	501.17	266.44	64.72	0.2429
LOEE (kWh/year)	296.70	4930.19	1873.37	813.61	0.4343
EIU	0.152 %	2.532 %	0.962 %	0.418 %	0.4343
FOI (int./year)	445	773	560.41	55.72	0.0994
DOI (h/int.)	0.291	0.705	0.470	0.0733	0.1560
ENSI (kWh/int.)	0.619	7.287	3.261	1.1635	0.3567

- 20 % increase of diesel fuel price (from 1.5 to 1.8 €/L).
- 40 % capital and replacement cost reduction of renewable energy technologies (WTs and PVs). This reduction may be attributed either to technology improvement and economies of scale, or to a modification in the regulatory regime that promotes the installation of RES technologies by offering incentives that reduce the capital and replacement cost of RES.

Tables 7–9 present the results of the above mentioned sensitivity analyses, as well as the initial scenario results for comparison purposes. More specifically, Table 7 presents the minimum *COE* values and their corresponding optimal configurations, Table 8 shows the results of the combined GA and local search procedure (referred to as GA-local search), and Table 9 shows the results of the MCS (average values). Regarding the comparison of GA-local search and MCS, the conclusions are similar with those mentioned in Sect. 5.3. Figure 6 shows the variability of *COE* obtained from all MCS compared to *COE* values obtained from the GA-local search procedure. From the study of Fig. 6 it can be concluded that: (i) all MCS obtained *COE* values are higher compared to those obtained from GA-local search procedure, (ii) the highest variability of the MCS results appears when considering residential CDFs (because of the exponential increase of residential customer damage cost with the increase of interruption duration), whereas the lowest variability

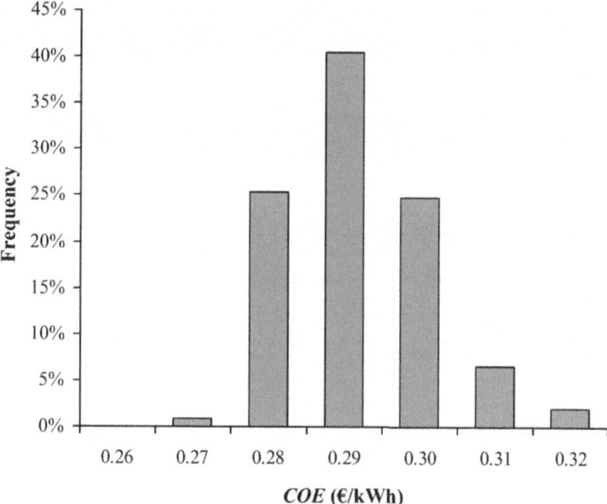

Fig. 4 *COE* histogram for agricultural load

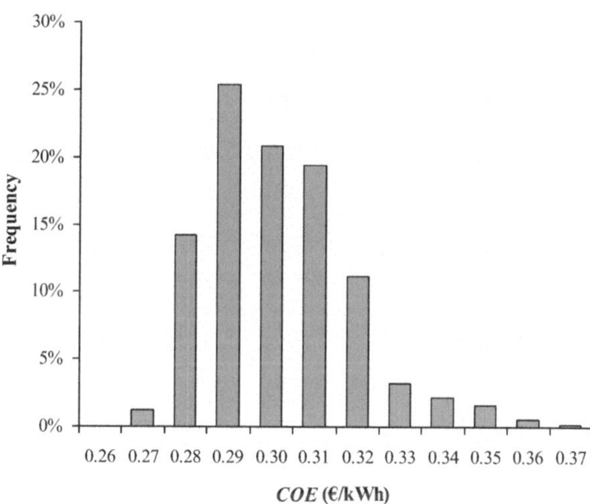

Fig. 5 *COE* histogram for residential load

appears when considering no customer damage cost, and (iii) in the majority of implemented MCSs, the average *COE* values assuming residential CDF are significantly higher compared to agricultural CDF.

The study of Tables 7–9 provides the following main conclusions for the considered case study system:

- The wind potential (scenarios 1 and 2) affects more the value of *COE* in comparison with the solar potential (scenarios 3 and 4).
- The optimal configurations of scenarios 3 and 4 (increased and decreased solar potential) are almost identical with the optimal configurations of the initial scenario.

Table 7 Optimal configuration for sensitivity analysis scenarios

Case	CDF	WTs	PVs (kW_p)	Dsl (kW)	Batteries	Converter (kW)	Dispatch strategy	*COE* (€/kWh)
Initial	No CDF	3	35	10	108	35	LF	0.2214
	Agricultural	3	50	40	144	40	LF	0.2659
	Residential	3	50	30	120	39	LF	0.2635
Wind +10%	No CDF	3	50	0	132	31	LF	0.2036
	Agricultural	3	42	40	120	39	LF	0.2466
	Residential	3	31	30	144	38	LF	0.2433
Wind −10%	No CDF	3	50	10	144	36	LF	0.2449
	Agricultural	4	50	40	144	41	LF	0.2918
	Residential	4	50	30	144	40	LF	0.2901
Solar +5%	No CDF	3	34	10	108	36	LF	0.2200
	Agricultural	3	50	40	132	40	LF	0.2627
	Residential	3	50	30	120	39	LF	0.2604
Solar −5%	No CDF	3	37	10	108	35	LF	0.2233
	Agricultural	3	50	40	156	40	LF	0.2693
	Residential	3	50	30	132	39	LF	0.2668
Diesel +20%	No CDF	3	39	10	96	37	LF	0.2287
	Agricultural	3	50	40	180	40	LF	0.2749
	Residential	3	50	30	168	40	LF	0.2721
RES −40%	No CDF	4	37	5	108	33	LF	0.1802
	Agricultural	4	50	40	108	41	LF	0.2197
	Residential	4	50	30	108	40	LF	0.2176

- The optimal configuration of scenario 1 (increased wind potential) considering no customer damage cost is the only case that does not contain the dispatchable diesel generator. As a result, the number of interruptions (FOI) is significantly increased.
- The (negative) effect of increased diesel fuel price (scenario 5) is marginally more severe than the (negative) effect of lower solar potential (scenario 4), but significantly less severe than the (negative) effect of lower wind potential (scenario 2).

Table 8 Sensitivity analysis results for GA—local search procedure

Case	CDF	GA—local search results						
		COE (€/kWh)	LOLE (h/year)	LOEE (kWh/year)	EIU (%)	FOI (int/year)	DOI (h/int)	ENSI (kWh/int)
Initial	No CDF	0.2214	1,053	9,708.73	4.987	689	1.529	14.091
	Agricultural	0.2659	2.50	6.22	0.003	13	0.192	0.478
	Residential	0.2635	55.50	224.92	0.116	148	0.375	1.520
Wind +10%	No CDF	0.2036	806	9,724.86	4.995	1,020	0.790	9.534
	Agricultural	0.2466	2.00	5.95	0.003	10	0.200	0.595
	Residential	0.2433	45.67	187.36	0.096	131	0.349	1.430
Wind −10%	No CDF	0.2449	1,044	9,708.60	4.987	650	1.606	14.936
	Agricultural	0.2918	3.50	7.65	0.004	16	0.219	0.478
	Residential	0.2901	56.83	237.55	0.122	147	0.387	1.616
Solar +5%	No CDF	0.2200	1,045	9,689.45	4.977	679	1.539	14.270
	Agricultural	0.2627	2.50	6.22	0.003	13	0.192	0.478
	Residential	0.2604	52.00	212.73	0.109	139	0.374	1.530
Solar −5%	No CDF	0.2233	1,055	9724.58	4.995	699	1.509	13.912
	Agricultural	0.2693	2.50	6.58	0.003	13	0.192	0.507
	Residential	0.2668	57.83	237.74	0.122	160	0.361	1.486
Diesel +20%	No CDF	0.2287	1,052	9,725.84	4.995	711	1.480	13.679
	Agricultural	0.2749	2.50	6.22	0.003	13	0.192	0.478
	Residential	0.2721	46.83	197.97	0.102	122	0.384	1.623
RES −40%	No CDF	0.1802	803	9680.81	4.972	519	1.546	18.653
	Agricultural	0.2197	2.17	5.93	0.003	11	0.197	0.539
	Residential	0.2176	35.83	149.76	0.077	94	0.381	1.593

- The lower cost of RES technologies (scenario 6) results in the system with the lowest cost (*COE*).
- Due to the minimum renewable fraction constraint value of 80%, all optimal configurations contain 3 to 4 WTs, whereas the PV installation is always greater that 30 kW_p, while in many cases the installed PV capacity is equal to its maximum possible value of 50 kW_p.

Table 9 Sensitivity analysis results for MCS

Case	CDF	MCS results (average values)						
		COE (€/kWh)	LOLE (h/year)	LOEE (kWh/year)	EIU (%)	FOI (int/year)	DOI (h/int)	ENSI (kWh/int)
Initial	No CDF	0.2286	1,303.61	11,850.02	6.086	798.80	1.634	14.850
	Agricultural	0.2867	210.16	1,820.96	0.935	566.40	0.364	3.098
	Residential	0.2965	266.44	1873.37	0.962	560.41	0.470	3.261
Wind +10%	No CDF	0.2057	1,067.53	11,591.58	5.954	1,357.88	0.786	8.537
	Agricultural	0.2648	214.11	1,852.61	0.952	582.52	0.360	3.056
	Residential	0.2879	312.79	2,390.13	1.228	610.26	0.507	3.826
Wind −10%	No CDF	0.2524	1,300.71	11,572.20	5.944	764.22	1.705	15.161
	Agricultural	0.3059	188.02	1604.03	0.824	488.63	0.379	3.193
	Residential	0.3155	221.96	1,566.89	0.805	436.75	0.502	3.497
Solar +5%	No CDF	0.2272	1,333.53	11,889.95	6.107	772.79	1.727	15.388
	Agricultural	0.2791	203.15	1726.46	0.887	556.09	0.360	3.011
	Residential	0.2920	267.11	1,952.32	1.003	540.73	0.489	3.525
Solar −5%	No CDF	0.2296	1,318.98	11,734.74	6.027	755.00	1.749	15.556
	Agricultural	0.2863	211.79	1,825.81	0.938	575.93	0.361	3.064
	Residential	0.3077	285.79	2101.94	1.080	586.30	0.479	3.458
Diesel +20%	No CDF	0.2360	1,297.18	11,773.18	6.047	780.61	1.664	15.096
	Agricultural	0.2946	194.82	1,807.78	0.929	489.78	0.389	3.538
	Residential	0.3077	251.56	2,042.88	1.049	455.23	0.544	4.348
RES −40%	No CDF	0.1859	994.18	10,839.14	5.567	877.58	1.133	12.357
	Agricultural	0.2307	132.06	963.78	0.495	411.33	0.317	2.283
	Residential	0.2367	206.00	1,372.09	0.705	420.37	0.485	3.189

- In all cases, when agricultural or residential CDFs are considered, the systems are notably more reliable (due to customer damage costs consideration). As a result, their optimal configuration contains significant higher capacity of the dispatchable diesel generator.
- In all the examined scenarios, the optimal configurations contain large number of batteries, converters of similar sizes, and adoption of LF dispatch strategy.

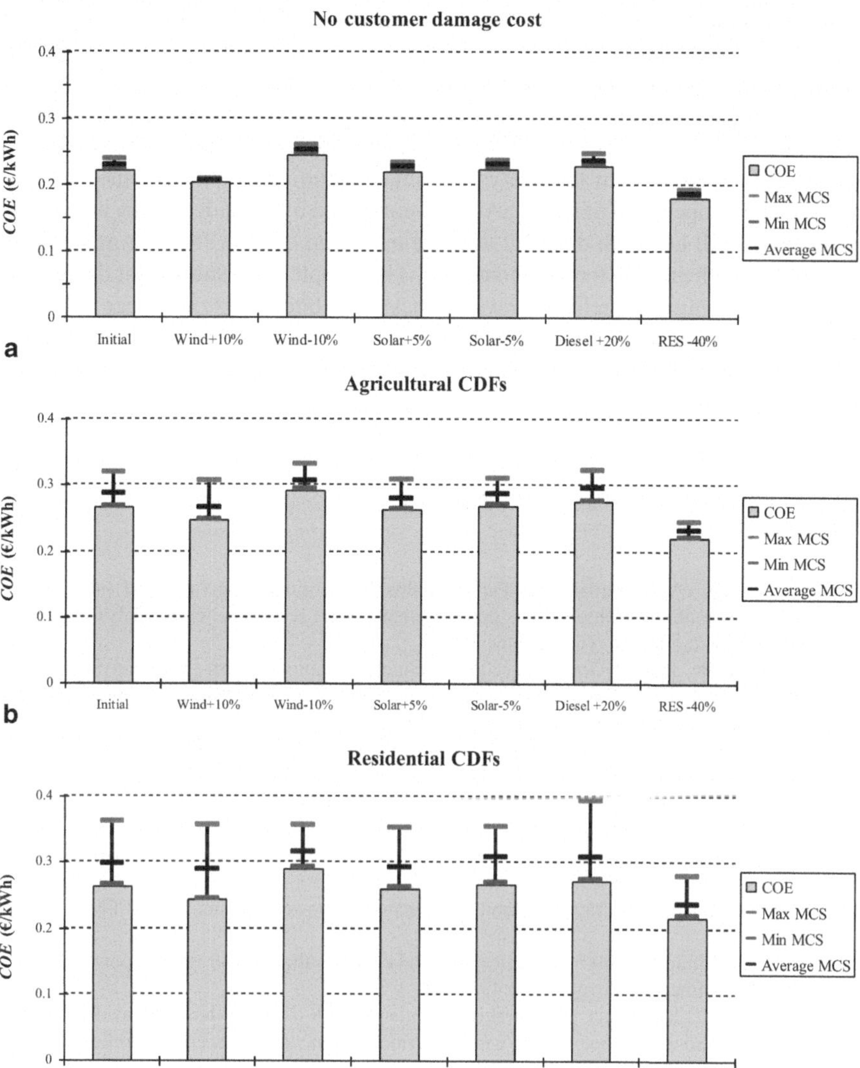

Fig. 6 Variability of obtained *COE* from MCS compared to GA-local search *COE* considering: **a** no customer damage cost, **b** agricultural CDF, and **c** residential CDF

7 Conclusions

The reliability evaluation of a SAPS that is based on renewable energy technologies is a complex and time consuming task, due to the intermittent nature of renewable resources, their variation, the high modularity of each part of the system, and the considered assumptions for the reliability analysis. In most cases, the optimal sizing procedure of such systems takes into account reliability issues in a generic framework, using general constraints (such as maximum unmet load constraint). However, in order to be complete, this analysis has to take into account the effect of two more parameters: the reliability worth as well as the forced outage rate of SAPS components. This chapter shows that the consideration of the reliability worth and the forced outage rate in the analysis changes significantly the obtained results. Moreover, the operation of a real SAPS, as computed by considering the above two parameters, will be much different than the operation of a SAPS ignoring both the reliability worth and the forced outage rate. This chapter also shows that the type of load, which changes the reliability worth, may also affect the performance of SAPS. The above conclusions have been drawn using sensitivity analysis considering a large number of alternative scenarios that take into account the uncertainty of weather and cost data.

References

1. Katsigiannis YA, Georgilakis PS, Papadopoulos SC, Moschakis MN (2012) Evaluating the performance of small autonomous power systems using reliability worth analysis. in Proc. PMAPS 2012, Istanbul, Turkey, June 2012
2. Kishore VVN, Jagu D, Gopal EN (2013) Technology choices for off-grid electrification. In: Bhattacharyya S. (ed) Rural electrification through decentralised off-grid systems in developing countries. Springer-Verlag, London, pp 39–72
3. Katsigiannis YA, Georgilakis PS, Tsinarakis GJ (2010) A novel colored fluid stochastic petri net simulation model for reliability evaluation of Wind/PV/Diesel small isolated power systems. IEEE Trans Syst Man and Cybern, Part A 40(6):1296–1309
4. Katsigiannis YA, Georgilakis PS (2013) Effect of customer worth of interrupted supply on the optimal design of small isolated power systems with increased renewable energy penetration. IET Gener Transm Distrib 7(3):265–275
5. Billinton R, Allan RN (1996) Reliability evaluation of power systems, 2nd ed. Plenum, New York
6. Karki R, Billinton R(2004) Cost effective wind energy utilization for reliable power supply." IEEE Trans Energy Convers 19(2):435–440
7. Koutroulis E, Kolokotsa D, Potirakis A, Kalaitzakis K (2006) Methodology for optimal sizing of stand-alone photovoltaic/wind-generator systems using genetic algorithms. Sol Energy 80:1072–1088
8. Katsigiannis YA, Georgilakis PS, Karapidakis ES (2010) Genetic algorithm solution to optimal sizing problem of small autonomous hybrid power systems. Lecture Notes in Artif Intell 6040:327–332
9. Katsigiannis YA, Georgilakis PS, Karapidakis ES (2010) Multiobjective genetic algorithm solution to the optimum economic and environmental performance problem of small autonomous hybrid power systems with renewables. IET Ren Power Gen 4:404–419
10. Georgilakis PS, Hatziargyriou ND (2013) Optimal distributed generation placement in power distribution networks: Models, methods, and future research. IEEE Trans. Power Syst 28(3)3420–3428
11. Katsigiannis YA, Georgilakis PS, Karapidakis ES (2012) Hybrid simulated annealing–tabu search method for optimal sizing of autonomous power systems with renewables. IEEE Trans Sustain Energy 3:330–338

12. Koskolos NC, Megaloconomos SM, Dialynas EN (1998)Assessment of power interruption costs for the industrial customers in Greece. in Proc. ICHQP '98, Athens, Greece, 1998
13. Manwell JF, McGowan JG, Rogers AL (2009) Wind energy explained, 2nd ed. Wiley, Chichester
14. Manwell JF, Rogers A, Hayman G, Avelar CT, McGowan JG (1998, November 2). http://ceere.ecs.umass.edu/rerl/projects/software/hybrid2/Hy2_theory_manual.pdf. Accessed 17 Feb 2014
15. Thomson M, Infield DG (2007) Impact of widespread photovoltaics generation on distribution systems. IET Renew Power Gener 1:33–40
16. Markvart T, Castañer L (2003) Practical handbook of photovoltaics: fundamentals and applications. Elsevier, UK
17. Skarstein O, Uhlen K (1989) Design considerations with respect to long-term diesel saving in wind/diesel plants. Wind Eng 13:72–87
18. Barley CD, Winn CB (1996) Optimal dispatch strategy in remote hybrid power systems. Sol Energy 58:165–179
19. Billinton R, Li W (1994) Reliability assessment of electric power systems using Monte Carlo methods. Plenum, New York

Condition Monitoring Benefit for Operation Support of Offshore Wind Turbines

Sebastian Thöns and David McMillan

1 Introduction

There is likely to be a large increase of installed capacity of offshore wind power in the coming decade. Assuming the European Union (EU) build rate during 2010 (883 MW installed [1]) can be increased to 1 GW per annum and sustained until 2020, and assuming an average capacity of 3.4 MW per turbine [2], this would result in ~3,500 wind turbine assets in the water. This is a conservative figure compared to some highly optimistic estimates, but is still a huge number of assets. Economic asset management of such a high number of units in such harsh environmental conditions, both in the short and long term, is non-trivial.

This chapter highlights the asset management challenges associated with this increased deployment by use of case studies, and proposes probabilistic methods to measure and reduce risk to investors and operators of offshore wind plant.

Offshore reliability and associated operation and maintenance cost estimation is an area of keen interest to wind farm operators. There is a high degree of uncertainty associated with these costs, coupled with some evidence showing O&M expenditures broadly increasing in early life [3]. To control this trend and to achieve risk reduction, models need to be developed in order to predict what is a neglected and important part of wind farm life cycle cost.

S. Thöns (✉)
Division 7.2 Buildings and Structures, BAM Federal Institute for Materials Research, Berlin, Germany
e-mail: sebastian.thoens@bam.de

D. McMillan
Institute for Energy & Environment, University of Strathclyde, Glasgow, UK
e-mail: dmcmillan@eee.strath.ac.uk

R. Karki et al. (eds.), *Reliability Modeling and Analysis of Smart Power Systems*, Reliable and Sustainable Electric Power and Energy Systems Management,
DOI 10.1007/978-81-322-1798-5_11, © Springer India 2014

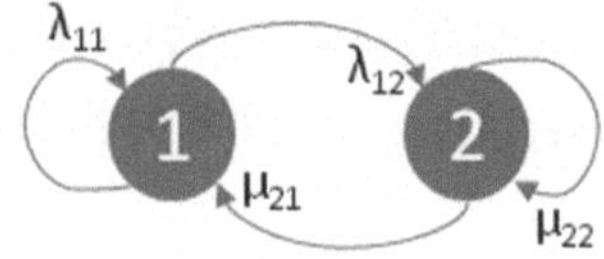

Fig. 1 Two state Markov chain

2 Operation Machinery Model

2.1 Short-Term Operation Machinery

Since operational information from offshore sites are sparse, the approach taken here is to examine data from onshore maintenance records and adjust the downtimes, lost energy, and failure rates to a level appropriate for offshore installations (there are precedents for this kind of approach such as [4]).

Previous studies showed how Markov chains coupled with Monte Carlo simulation provide a suitably flexible approach for modeling wind farm reliability and O&M [5, 6]. The methodology has been successfully adopted by several other authors [7, 8] to solve similar problems. In this chapter, the approach is utilized to produce a cost-benefit analysis of condition based maintenance.

2.2 Failure Modeling

Following a reliability centered maintenance study (see e.g., [9]), failure rates for a set of wind turbines have been derived. Individual asset groups are then modeled by a Markov chain. For the simplest case, consider Fig. 1, a two-state chain where state 1 is operational and state 2 is failed, failure rate of component is λ_{12} and repair rate is μ_{21}.

The probability of remaining in each state during time step Δt:

$$\lambda_{11}\Delta t = 1 - \lambda_{12}\Delta t \tag{1}$$

$$\mu_{22}\Delta t = 1 - \mu_{21}\Delta t \tag{2}$$

Probability of being in state 1 after time step Δt:

$$P_1(t+\Delta t) = P_1(t)\lambda_{11}\Delta t + P_2(t)\mu_{21}\Delta t \tag{3}$$

$$P_1(t)' = -\lambda_{12}P_1(t) + \mu_{21}P_2(t) \tag{4}$$

System equations such as Eq. (4) can be solved algebraically for simple systems, or discretised for the purposes of Monte Carlo simulation for complex systems where multiple constraints, such as weather access windows, can be easily modeled.

Table 1 Model cost assumptions. MTTR is based on major failures and assumes time-based maintenance

Failure mode	$\lambda(\Delta t=1$ year)	MTTR (days)	P(minor)	P(major)	C(minor) €	C(major) €
Controller	0.176	1	0.450	0.550	1,995	16,991
Gearbox	0.180	7	0.978	0.022	16,706	4,89,768
Nacelle	0.328	1	0.536	0.464	1,093	2,749
Transmission	0.035	43	0.000	1.000	N/A	2,34,098

Intermediate states can be used for plant items which degrade slowly over time, such as bearings.

The asset categories used in this study differ from previous studies, which have focused on gearbox, electronics, generator, and rotor [5]. These components have been revised on the basis of recent reliability centered maintenance (RCM) studies, and are summarized in Table 1. The fault occurrence rate, θ, is derived on the basis of a utilities asset management system, which comprised 84 turbines and 255 operation years. The turbines are modern multi-MW machines within the range 3–5 years of operation. A fault occurrence is classed as anything that causes the wind turbine to stop functioning, no matter how trivial. Thus the database encompasses all failure events: from those requiring a very short maintenance visit, to those requiring cranage, additional specialist labor, and large component replacement cost. The top four components are modeled in this study, however, tower is replaced with gearbox owing to the very low-impact nature of most tower faults, which mostly relate to maintenance access systems.

2.3 *Cost Modeling and Assumptions*

By analyzing maintenance databases, it is possible to extract many useful metrics which can be used for populating a maintenance cost model. The failure rate (λ)—a subset of the occurrence rate (θ) shown in Table 2—and mean time to repair (MTTR) are the most obvious metrics. In addition, it is possible to study the severity of faults and their likelihood. In this chapter, we consider minor and major failures, with associated probability (P(minor), P(major)) and cost (C(minor), C(major)). These are shown in Table 1.

Modeling of time-based maintenance (TBM) is based on restoration of the Markov chain to fully operating condition once per annum. This incurs minor costs as shown in Table 1. However, in the event of an unplanned failure, a cost premium of 50 % is applied to the incurred costs. This is broadly representative of specialized vessel hire and labor at short notice, and the possible need for fast fabrication and shipment of components, again in an unplanned, expedited manner.

The key assumptions underpinning the condition-based maintenance (CBM) model are that via better maintenance planning, costs are kept to the values shown in Table 1. In addition, the MTTR for a gearbox is reduced to 3 days and transmission to 7 days. Since the chief operational advantage of CBM is increased scope for

Table 2 Fault occurrence rate by asset group—onshore data [10]

Asset name	θ (Δt = 1 year)
Controller	2.362
Nacelle	1.391
Tower	1.221
Transmission	1.091
Gearbox	0.841
Hub	0.490
Parking brake	0.380
Hydraulics	0.360
Yaw	0.270
Generator	0.230
Pitch	0.220
Measurement (sensors, etc)	0.210
Blade system	0.060
Switchgear	0.060
Over speed protection system	0.040
HV system	0.040

Table 3 Operational constraints

Vehicle	Wave climate influence	Typical access limit (m)
Access transfer boat	Operability limited by wave height	1.5
Field support vessel	Limit depending on turbine access method	1.5–3
Helicopter	Limited by health and safety for sea rescue	3.5
Mobile jack-up vessel	Jack-up and movement limited by sea state	2

planning, it is appropriate that modeling of CBM explicitly captures the effects of improved planning to reduce downtime and procure in a planned, low-risk manner.

The energy yield model [5] is based on wind data simulated from a coastal location in the UK, as offshore data were not available [11]. The equivalent capacity factor is 35 %. The main cost assumption is that the electricity production credit is € 126/MWh. This is based on future reforms to the UK ROC system and is equivalent to 1.8 ROCs/MWh.

2.4 Constraint Models

Significant wave height is the dominant access constraint for maintenance of offshore wind farms, as large waves prevent crew transfer as well as jack-up and heavy lift operations. The importance of access constraints in operational terms is clearly illustrated by Table 3 (see Dinwoodie et al. [12] for further discussion). The vast majority of maintenance visits will involve crew transfer only.

In order to apply access constraints it is required to build a significant wave height time series which adequately captures the statistical properties of the original

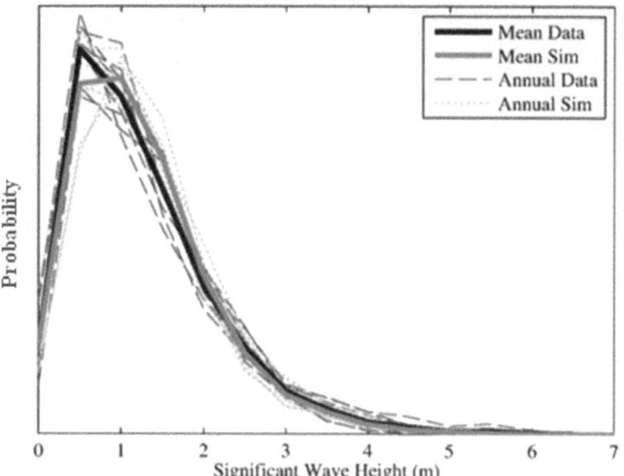

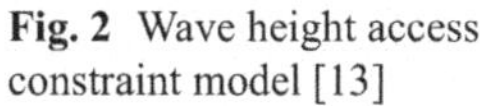
Fig. 2 Wave height access constraint model [13]

series such as annual probability distribution, short-term correlation structure, and seasonality. The time series utilized in this chapter is based on a probabilistic wave height time series model developed in [13] and [14]. Figure 2 clearly shows that the model captures the annual probability distribution. Other important aspects of the model are discussed in [14].

At the moment, crew transfer to offshore wind farms is generally constrained at wave heights of 1.5 m or over. Thus all modeled maintenance actions (CBM, TBM, and unscheduled) are affected by this constraint. Future work will consider the impact of access constraints for different operations such as jack-up and heavy lift.

2.5 *Case Study*

In the case study we consider a 5 MW offshore machine with failure rates and MTTR as shown in Table 1. Multiple simulations are run at a time resolution of 1 day. Wind speed, energy yield, revenue generation, incurred O&M cost, and weather constraint are all factored in as explained in previous sections. Since failure rates are expected to increase in the offshore environment, we increase λ and perform a sensitivity study to evaluate the two maintenance methods. The analysis is carried out for two cases:

- *Case 1* (optimal CBM)—all four asset failure modes are subject to CBM (operating with reduced MTTRs and base costs described in Sect. 2.3)
- *Case 2* (realistic CBM)—gearbox and transmission are subject to CBM, but MTTRs are increased to 5 and 20 days respectively. For TBM, no cost premium is applied to nacelle and controller. Repair cost premiums for gearbox and transmission are inflated by only 10 % instead of the 50 % in case 1.

Figure 3 shows the impact on availability as λ is increased. It can be seen that despite the increase in λ, the CBM policy maintains availability at a much higher level.

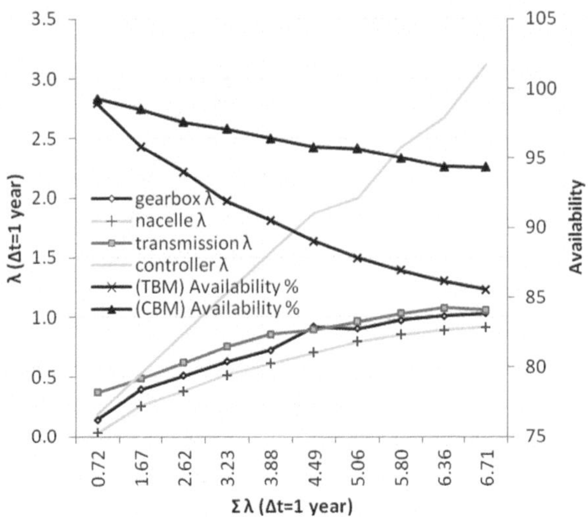

Fig. 3 Case 1. Impact of increased failure rates on asset availability under time based and condition based maintenance

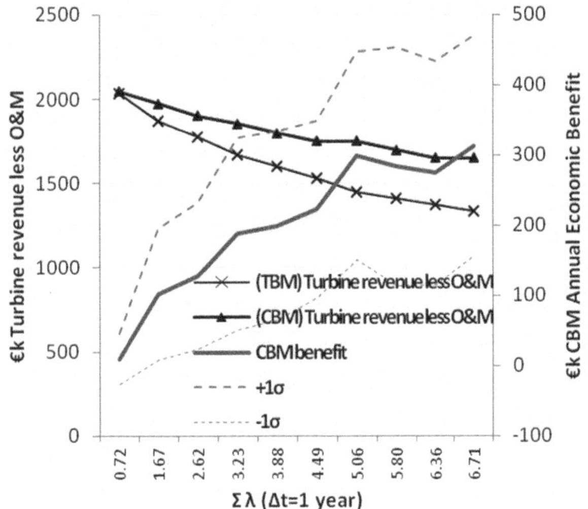

Fig. 4 Case 1. Economic benefit of condition based maintenance under model assumptions

Failure initiations are increased for all four failure modes at the same rate. Three failure modes are mitigated by TBM or CBM, however, controller failures cannot be mitigated via maintenance and hence show the biggest proportional increase in Fig. 3. This illustrates the potential technical benefits of CBM, and the potential importance of fault tolerance in power electronic design [15, 16]. Interestingly, the results from the TBM policy are in the same ballpark as availability figures from round 1 offshore sites in the UK [13]. Figure 4 shows an economic benchmark between TBM and CBM. It is noted that for the starting values of λ (see Table 2), the CBM policy is approximately at economic parity with TBM. Only when failures begin to increase is the value of the CBM policy obvious.

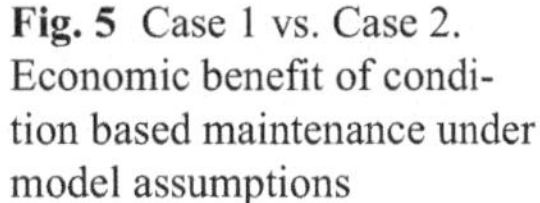

Fig. 5 Case 1 vs. Case 2. Economic benefit of condition based maintenance under model assumptions

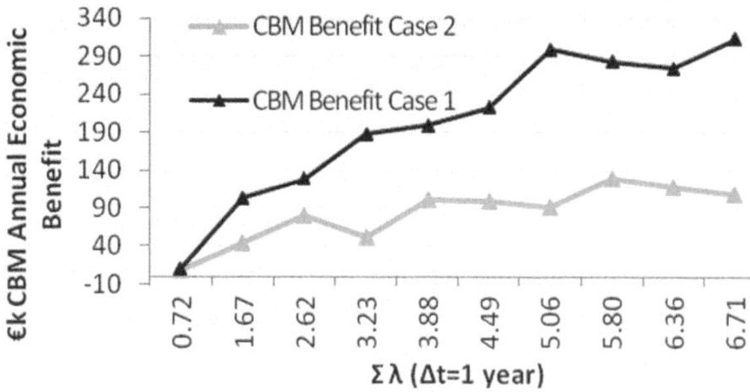

Figure 5 illustrates how a more realistic representation of the savings to be expected from CBM (Case 2) significantly alters the economic benefit, implying a need to understand how decision processes such as maintenance scheduling, accessibility, and spares provision can be supported by data subject to real world constraints.

Reliability of future offshore sites remains to be seen, however, UK round 1 sites had an average availability of roughly 80 %. This implies a failure rate towards the top end of the spectrum of those considered in this chapter (see Fig. 3), suggesting that CBM will be required to reduce costs. This economic benefit will depend upon how well this data can be used to support operational decisions. In the longer term the economic benefit of CBM lies in managing asset structural integrity and life extension.

3 Structural Risk and Integrity Management

The basic idea of the structural risk and integrity management is the cost-efficient mitigation of structural risks for securing the functioning of the support structure throughout the life cycle. Structural risks are characterized by low probabilities of failure but high consequences such as the loss of one plant or a wind park. The structural integrity management comprises the inspection and maintenance in combination the monitoring of the wind turbine support structure.

The management of the structural integrity of a support structure constitutes in practice a task of the operator of a wind park who is subjected to the inspection and maintenance handbook of the designer as well as the structural code and regulation requirements such as [17, 18]. These requirements leave limited room for optimization and thus the basis of the codes and regulations namely a life cycle cost-benefit analysis in the operation phase is the starting point for the introduced approach.

On the basis of an optimization of a life cycle cost-benefit analysis comprising the inspection, maintenance and repair costs, the failure costs, and the costs of human safety usually the target reliabilities for the structures are determined (e.g., [19, 20]). This optimization of the life cycle costs of a structure accounts for the boundaries in the context of the present code generation (see e.g., the Linds postulate, e.g., [20, 21]).

The target reliabilities (or probability of failure thresholds) can then be compared to the results of the structural condition assessment and can serve as a basis for the determination of the inspection intervals (e.g., [20, 22, 23]).

The optimization of the life cycle costs must not be restricted only to the reliability level but can include further decision variables. Recently it has been shown that monitoring systems can significantly influence the expected life cycle cost of an offshore structure implying that monitoring systems give more certain information about the condition of the structure [24]. Furthermore, the monitoring of the wind turbine and its structure is already a part of the regulation applying to an offshore wind park in the external economy zone in Germany [17, 18]).

The aim of the following sections is to analyze the influence of monitoring systems first on the expected failure costs, i.e., on the risks, and second on the expected costs of the structural integrity management. For this aim the fundamentals of a life cycle cost-benefit analysis are outlined in the section 3.1 and an expected monitoring benefit related to the failure costs and to the structural integrity management costs are derived. In section 3.2 the parameters, i.e., the decision variables, to be considered are derived and the optimization aims are formulated. Section 3.3 contains then the outline of a case study and the results.

3.1 Long-Term Structural Operation Model

The long-term structural operation model comprises the expected value of failure costs $E[C_F]$, the expected costs of the structural integrity management $E[C_{SIM}]$ which itself consists of the expected inspection costs $E[C_I]$ and the expected repair costs $E[C_R]$ for the determination of the expected total life-cycle costs $E[C_T]$ (Eqs. (5) and (6), see e.g., [22]).

$$E\left[C_T\right] = E\left[C_F\right] + E\left[C_{SIM}\right] \tag{5}$$

$$E\left[C_{SIM}\right] = E\left[C_I\right] + E\left[C_R\right] \tag{6}$$

Such a life-cycle cost-benefit analysis involves various probabilistic models. The individual structural events no failure, failure, inspection, and repair of a component are modeled with a decision tree for each service year in the life cycle. The probabilities of the events no failure and failure are calculated with a structural system reliability analysis which requires probabilistic loading, structural, and limit state models. The inspection events are calculated on the basis of reliability based inspection planning which itself builds upon a structural reliability analysis and requires an inspection and repair strategy as well as the definition of a target probability of failure. The consequence model accounts for the costs of the structural events. The structural events described by their probabilities and consequences are modeled with a decision tree throughout the life cycle of the support structure.

It has been demonstrated in [24] that the reliability calculated with monitoring data and its associated models can be higher in comparison with the design data and models. This increase in reliability is caused by lower (model) uncertainties for the utilization of the monitoring data for the case of low measurement uncertainties. By the change, i.e., the increase, of the structural reliability, the expected total life cycle costs are affected and Eq. (5) is rewritten to Eq. (7). The expected costs of failure $E\left[C_F^M\right]$ additionally include now the costs associated to the loss of monitoring system.

$$E\left[C_T^M\right]=E\left[C_F^M\right]+E\left[C_{SIM}^M\right] \tag{7}$$

The expected costs for the structural integrity management including monitoring $E\left[C_{SIM}^M\right]$ comprise, beside the expected inspection and repair costs, the expected and channel k dependent costs of the monitoring system $E\left[C_{Sys}^M(k)\right]$ and its installation $E\left[C_{Inst}^M(k)\right]$ as well as the monitoring system operation $E\left[C_{Op}^M\right]$ (Eq. (8)). The operation costs are discounted with the discount rate i_r to the present value dependent on the time of cash flow t and are multiplied by the probability of no failure $(1-p_F)$ (Eq. (9)).

$$E\left[C_{SIM}^M\right]=E\left[C_I^M\right]+E\left[C_R^M\right]+E\left[C_{Sys}^M(k)\right]+E\left[C_{Inst}^M(k)\right]+E\left[C_{Op}^M\right] \tag{8}$$

$$E\left[C_{Op}^M\right]=(1-p_F)\cdot C_{Op}^M\cdot\frac{1}{\left(1-i_r\right)^t} \tag{9}$$

The expected failure costs and the expected costs for the structural integrity management can be calculated with this approach for both cases, namely a structure without a monitoring system and a structure with a monitoring system. Furthermore, an expected monitoring benefit $E[B_M]$ as the difference of the expected costs with and without monitoring ($E[C]$ and $E[C^M]$) can be calculated (Eq. (10)).

$$E\left[B_M\right]=E\left[C_T\right]-E\left[C_T^M\right] \tag{10}$$

3.2 Decision Variables and Optimization Aims

The long-term structural operation model is applied with the framework of Bayesian preposterior decision analysis facilitating to account for yet unknown (monitoring) information and the formulation of decision variables and optimization aims.

Basically, the application of a monitoring system involves the decision where to monitor and how many components to monitor. This generic decision can be formally written with Eq. (11) and where D_1 denotes the decision set consisting of n different component sets c_{Si}.

$$D_1 = \{c_{S1}, c_{S2}, \ldots, c_{Sn}\}, \tag{11}$$

Furthermore, it has to be accounted for a criterion modeling the performance of the monitoring system in a structural reliability analysis. This criterion constitutes the reduction of the probability of failure by the monitoring data Δp_f^M (Eq. (12)), see [24].

$$D_2 = \{\Delta p_{f,1}^M, \Delta p_{f,2}^M, \ldots, \Delta p_{f,n}^M\}, \tag{12}$$

Two different optimization aims can be formulated with Eq. (10). The first optimization aim is to maximize the benefit related to the expected failure costs, i.e., to the risks, accounting for the decision variables (Eq. (13)). What is written here in the form of an equation constitutes the common association to the purpose of monitoring, namely that monitoring reduces the risks. Given that association, the expected monitoring benefit related to the failure costs should be always positive.

$$E\left[B_{M,C_F}\left(c_S, \Delta p_f^M\right)\right] = \arg\max\left(E\left[C_F\left(c_S, \Delta p_f^M\right)\right] - E\left[C_F^M\left(c_S, \Delta p_f^M\right)\right]\right) \tag{13}$$

The second aim, most interesting for an operator, is the monitoring benefit caused by the difference of the expected structural integrity management costs $E\left[B_{M,SIM}\right]$ (Eq. (14)).

$$E\left[B_{M,SIM}\left(c_S, \Delta p_f^M\right)\right] = \arg\max\left(E\left[C_{SIM}\left(c_S, \Delta p_f^M\right)\right] - E\left[C_{SIM}^M\left(c_S, \Delta p_f^M\right)\right]\right) \tag{14}$$

3.3 Case Study

The cost-benefit analysis model introduced in the preceding section is now applied to the reference case which constitutes a tripod support structure of a Multibrid M5000 prototype offshore wind turbine. The reliability analysis and the results are documented in [25] comprising 92 hot spots of the tower segments, the braces, central tube, and the pile guides of a tripod for the considered fatigue limit state.

The inspection strategy builds upon the approaches of the reliability inspection planning and is based on the models in [22]. Target probabilities of failure of 1.00×10^{-4} and 1.00×10^{-3} are assumed. A magnetic particle inspection (MPI) is assumed as the inspection technology (see [22, 26]). The repair event is defined on the basis of a measured crack size during an inspection (see [22]).

The consequence model consists of failure costs $C_F = 1$, inspection costs $C_I = 10^{-3}$ and repair costs $C_R = 10^{-2}$ per component and an interest rate of $i_r = 5\%$

Table 4 Example of the cost model associated with the reference case

Type of costs	Value
Failure costs C_F	7,500,000 €
Inspection costs per component C_I	7,500 €
Repair costs per component C_R	75,000 €
Costs of monitoring system p. channel $C^M_{Sys}(k)$	1,000 €/k
Costs of system installation p. channel $C^M_{Inst}(k)$	1,000 €/k
Cost of system operation per year C^M_{Op}	5,000 €/a

which represent generic assumptions [22]. In relation to this cost model, a monitoring cost model for the reference case is introduced. The costs of the monitoring system are assumed to $C^M_{Sys}(k) = 1.33 \cdot 10^{-4}$ per channel, where three channels (i.e. sensors) are associated with the monitoring of one hot spot. The costs of installation are assumed to $C^M_{Inst}(k) = 1.33 \cdot 10^{-4}$ per channel and the operation costs are assumed to $C^M_{Op} = 6.67 \cdot 10^{-4}$ per year. As an example for the cost model the reference case is considered assuming generic costs of 1,500,000 € per MW [27]. The resulting costs for the reference case are summarized in Table 4; the analysis is performed with the normalized cost model as described. Further, a yearly probability of failure threshold of 1.00×10^{-3} and of 1.00×10^{-4} is considered.

The expected monitoring benefits are calculated utilizing the structural system reliability analysis [25], the influence of the monitoring data on the individual component reliabilities [24], the reliability based inspection, and repair planning model as described.

The calculated monitoring benefits are depicted in Figs. 6 and 7. Two probabilities of failure reduction factors Δp^M_f of 2.0 and 3.0 (left/right in Figs. 6 and 7) and two yearly probabilities of failure thresholds of 1.00×10^{-3} and of 1.00×10^{-4} are considered (Figs. 6 and 7).

For both probabilities of failure thresholds and both probabilities of failure reduction factors the expected failure cost benefit is positive for all number of monitored components. The higher the number of monitored components the higher failure cost benefit, i.e., the lower the risks associated to a structure with a monitoring system, until a maximum of monitored components is reached.

The behavior of the expected costs of the structural integrity management (dashed lines) is more complex. For a yearly probability of failure thresholds of 1.00×10^{-3} the benefit is negative with minor dependency on the number of monitored components. For a yearly probability of failure thresholds of 1.00×10^{-4} the benefit becomes positive with 7 monitored components and the probability of failure reduction factor of 3.0. The expected benefit is then increasing until a number of 19 monitored components. It turns out that 19 hot spots of the support structure are subjected to the reliability based inspection, i.e., that 19 hot spots have to be inspected during the service life of the structure.

On the basis of the considered examples, an optimum monitored component set of $D_{1,opt} = \{c_{S,5} = 5\}$ for a target probability of failure of 1.00×10^{-3} and of $D_{1,opt} = \{c_{S,19} = 19\}$ for a target probability of failure of 1.00×10^{-4} are determined

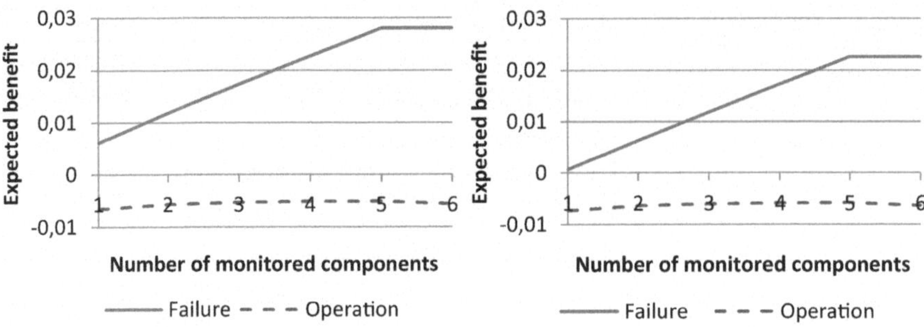

Fig. 6 Expected monitoring benefits $E\left[B_{M,C_F}\right]$ and $E\left[B_{M,SIM}\right]$ (*dashed*) for yearly probability of failure thresholds of 1.00×10^{-3} and for a probability of failure reduction factor Δp_f^M of 2.0 (*left*) and 3.0 (*right*)

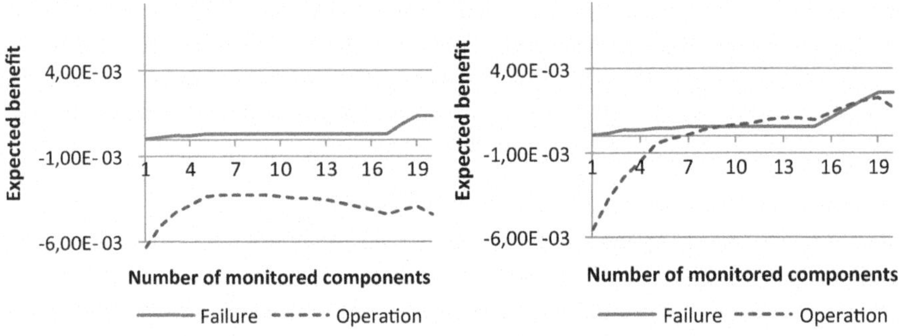

Fig. 7 Expected monitoring benefits $E\left[B_{M,C_F}\right]$ and $E\left[B_{M,SIM}\right]$ (*dashed*) for yearly probability of failure thresholds of 1.00×10^{-4} and for a probability of failure reduction factor Δp_f^M of 2.0 (*left*) and 3.0 (*right*)

(Fig. 8). The optimum probability of failure reduction factor which is dependent on the monitoring and measurement uncertainty is determined as 3.0 in this exemplary study. Considering all produced results (see [24]) the optimum constitutes the maximum achievable uncertainty reduction factor $D_{2,opt} = \max\left\{\Delta p_f^M\right\}$.

4 Summary and Conclusions

This chapter contains actual research results in the field of condition monitoring support for the operation of offshore wind turbines. Both the machinery operation and the structural integrity management are addressed. It can be concluded that monitoring systems can support the operation management by reducing costs and risks.

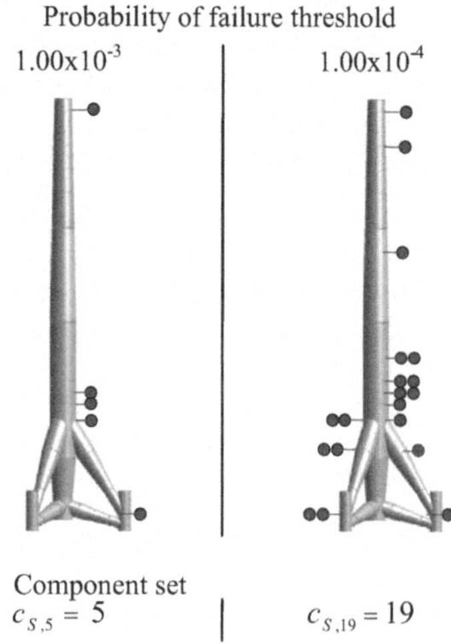

Fig. 8 Optimal monitoring component sets for a tripod support structure of a Multibrid M5000 prototype offshore wind turbine in dependency of the probability of failure thresholds of 1.00×10^{-3} (*left*) and of 1.00×10^{-4} (*right*)

For the machinery operation it is shown that CBM has significant cost advantages for the expected offshore failure rates. Clearly, cost-effective CBM requires a reliable monitoring system, and the CM information must be utilized by O&M planners to reduce MTTR and plan spares procurement in an efficient manner. The integrity management of the support structure can be supported by monitoring systems building upon a long-term structural operation analysis explicitly accounting for monitoring and its results within the framework of the Bayesian decision analysis. With a case study considering a Multibrid M5000 wind turbine structure, it is shown how a monitoring system can be optimized for risk reduction and/or for the reduction of the expected structural integrity management costs.

This chapter is seen as the first step in developing holistic monitoring systems and approaches for the support of the offshore wind turbine operation.

References

1. EWEA (2011) The European offshore wind industry key trends and statistics 2010, January 2011
2. EWEA (2011) The European offshore wind industry—key trends and statistics: 1st half 2011. http://www.ewea.org/fileadmin/ewea_documents/documents/00_POLICY_document/Offshore_Statistics/20112707OffshoreStats.pdf. Accessed 13 Feb 2014
3. Ernst&Young (2009) Cost of and financial support for offshore wind. Department of energy and climate change
4. van Bussel GJW, Zaaijer MB (2001) Reliability, availability and maintenance aspects of large-scale offshore wind farms, a concepts study. Delft University of Technology, Delft
5. McMillan D, Ault GW (2008) Condition monitoring benefit for onshore wind turbines: sensitivity to operational parameters. IET Renew Power Gen 2:60–72
6. McMillan D, Ault GW (2010) Techno-economic comparison of operational aspects for direct drive and gearbox-driven wind turbines. IEEE Trans Energy Conver 25:191–198
7. Besnard F, Bertling L (2010) An approach for condition-based maintenance optimization applied to wind turbine blades. IEEE Trans Sustain Energy 1:77–83

8. Byon E, Ntaimo L, Ding Y (2010) Optimal maintenance strategies for wind turbine systems under stochastic weather conditions. IEEE Trans Rel 59:393–404
9. Guevara Carazas FJ, Martha de Souza GF (2009) Availability analysis of gas turbines used in power plants. Int J Thermodyn 12(1):28
10. McMillan D, Ault G (2012) Towards reliability centred maintenance of wind turbines, presented at the probabilistic methods applied to power systems. Istanbul 2012
11. Hill D, Bell K, Mcmillan D, Infield D (2012) Application of auto-regressive models to UK wind speed data for power system impact studies. IEEE Trans Sustain Energy 3:8
12. Dinwoodie I, Catterson V, McMillan D (2013) Wave height forecasting to improve offshore access and maintenance scheduling, presented at the IEEE PES general meeting 2013, Vancouver, Canada
13. Dinwoodie I, McMillan D (2012) Analysis of offshore wind turbine operation & maintenance using a novel time domain meteo-ocean modeling approach, presented at the ASME Turbo Expo Copenhagen, Denmark
14. Dinwoodie IA, McMillan D, Quail F (2012) Sensitivity of offshore wind turbine operation & maintenance costs to key operational parameters, presented at the European safety reliability and data analysis conference ESReDA 2012, Glasgow, UK
15. Zhang CW, Zhang T, Chen N, Jin T (2013) Reliability modeling and analysis for a novel design of modular converter system of wind turbines. Reliab Eng Syst Saf 111:86–94
16. Parker MA, Ng Chong, Ran Li (2011) Fault-tolerant control for a modular generator–converter scheme for direct-drive wind turbines. IEEE Trans Ind Electron 58:305–315
17. Konstruktive Ausführung von Offshore-Windenergieanlagen, 2007
18. Richtlinie für die Zertifizierung von Condition Monitoring Systemen für Windenergieanlagen, GL Wind IV—Teil 4, 2007
19. Straub D (2004) Generic approaches to risk based inspection planning for steel structures. ETH Zürich, Zürich
20. JCSS (2001) Probabilistic assessment of existing structures. A publication of the Joint Committee on Structural Safety. RILEM Publications S.A.R.L
21. Lind NC (1978) Reliability-based structural codes. Optimization theory; reliability-based structural codes. Practical calibration; safety level decisions and socio-economic optimization. In Safety of structures under dynamic loading. vol. 1, Tapir, Trondheim, pp. 135–175
22. Straub D (2004) Generic approaches to risk based inspection planning for steel structures. PhD. thesis, Chair of Risk and Safety, Institute of Structural Engineering, ETH Zürich, Zürich
23. Probabilistic Model Code, 2006
24. Thöns S (2011) Monitoring based condition assessment of offshore wind turbine structures. Dr. sc. (ETH Zürich) PhD thesis, Chair of Risk and Safety, Institute of Structural Engineering, ETH Zurich, Zurich
25. Thöns S, Faber MH, Rücker W (2010) Support structure reliability of offshore wind turbines utilizing an adaptive response surface method. In 29th international conference on ocean, offshore and Arctic engineering (OMAE 2010)
26. Visser Consultancy Limited (2000) POD/POS curves for non-destructive examination
27. European Wind Energy Association (EWEA) (2009) The economics of wind energy

Towards Reliability Centred Maintenance of Wind Turbines

David McMillan and Graham W. Ault

1 Introduction

Control of operation and maintenance (O&M) cost is an area of growing interest to wind farm operators, as groups of assets come to the end of equipment manufacturers warranty agreements. Reliability-centred maintenance (RCM) has very successful previous applications in thermal plant, and is applied here to wind turbines. By identifying key components and quantifying risk, maintenance effort can be focused on appropriate areas thus lowering O&M spend in the longer term.

2 Previous Work and Literature Review

2.1 RCM and Wind Plant

Ribrant and Bertling [1] carried out a study which provided comprehensive failure rate and downtime data by WT subassembly. The database comprised many different WT models and manufacturers. This chapter also contains a study of gearbox failure modes, including repair and replacement statistics. Such detailed failure information provides an important and rare insight into wind farm operational issues.

Rademakers et al. [2] looked at a structural breakdown of parts within a wind turbine and discussed failure detection methods such as inspection and condition monitoring. A fault tree analysis was carried out for the component parts such as rotor, nacelle and tower. From this detailed analysis a flaw in the design of the studied turbine was detected and the authors suggested more sensor redundancy to

D. McMillan (✉) · G. W. Ault
Institute for Energy & Environment, University of Strathclyde, Glasgow, UK
e-mail: dmcmillan@eee.strath.ac.uk

G. W. Ault
e-mail: g_ault@eee.strath.ac.uk

R. Karki et al. (eds.), *Reliability Modeling and Analysis of Smart Power Systems*, Reliable and Sustainable Electric Power and Energy Systems Management,
DOI 10.1007/978-81-322-1798-5_12,

cut down the risk of failure, showing the value of such an approach. Similar studies were carried out by Michos et al. [3] with the focus on safety issues. More recently, Arabian-Hoseynabadi et al. [4] describe the failure modes and effects analysis (FMEA) approach and apply it to a set of WT reliability data.

Andrawus et al. [5] examined RCM as part of maintenance modeling. Data from operational wind farms are used to populate the model in order to establish an optimal maintenance policy. Similar studies can be found in [6–8] which deal with different aspects of the problem such as specific component parts, implementation of condition based maintenance (CbM) or the effect of seasonal weather patterns.

This chapter contains a real application of RCM methods to a fleet of operational wind farms. By analysing operational maintenance data, important failure modes can be highlighted and action taken to mitigate them. Such methods are underutilised in the wind industry at present.

2.2 RCM Literature

Several excellent RCM resources are available, such as [9, 10]. A comprehensive and transparent case study, where RCM is applied to gas turbine power plant, can be found in [11]—in this case the authors importantly measure the plant performance improvement after one iteration of the RCM process. Such performance benchmarking is crucial in the longer term to justify the resource allocation to RCM, as well as any extra labour or material cost incurred by mitigation methods such as condition monitoring.

3 RCM Process

The RCM process can be broken down into the following steps [9, 11], which are adopted in this chapter:

Step 1: System selection and information collection

- Collect data.
- Define system boundaries.

Step 2: Develop understanding of system

- Define sub systems.
- Functional tree.

Step 3: Define system functions and functional failures

- FMECA.
- Data processing.
- Detailed risk analysis (main result).

Table 1 Maintenance data summary

Site	MW	Model	$\#WT_n$	Data start	Data end	Months	Δt_n(years)
Site 1	36.8	A	16	11/2008	02/2011	28	2.33
Site 2	119.6	B	52	11/2007	02/2011	40	3.33
Site 3	36.8	A	16	05/2008	11/2010	31	2.83

Step 4: Task selection (feedback)

- Identify components for more maintenance effort.
- 'Prioritisation' of maintenance based on criticality.
- Is condition monitoring justified?

3.1 Data Sources

The main source of data used in the analysis was a set of maintenance records used as part of a maintenance management system [12]. Table 1 illustrates the data set available, where $\#WT_n$ is the number of wind turbines at site n and Δt_n is the time in years covering the maintenance record from that site. WT models A and B are of similar design and have a large majority of common components.

It can be seen from the data start and data end fields in Table 1, parallel streams of data are available from more than one site for some periods, whereas for other time periods there is no coverage.

In order to calculate annual occurrence and failure rates (λ), the total number of WT operational years $WT\Delta t_{total}$ must be deduced using Eq. (1). This is calculated in Table 1 as 255.72 WT-years equivalent. This figure is used to calculate failure rates Eq. (2) for subcomponents common to WT models A and B, where components are exclusive to model A or B, then $WT\Delta t_A = 82.56$ (WT-years equivalent) and $WT\Delta t_B = 173.16$.

$$WT\Delta t_{total} = \sum_{n=1}^{3} \Delta t_n \times \#WT_n \tag{1}$$

$$\lambda_{FM\#} = \frac{Freq \cdot FM\#}{WT\Delta t_{total}} \tag{2}$$

3.2 System Description

The system under study is a Danish concept multi-MW onshore wind turbine. Analysis of the system was limited to the wind turbine asset and switchgear—inter array transmission was not included. Asset sub-groups were defined via an existing wind farm operators asset structure and were allocated a failure mode number ($FM\#$), as shown in Table 2.

Table 2 Asset group failure mode number

Asset group	*FM#*
Overall asset	0
Blade system	100
Parking brake	200
Controller	300
Gearbox	400
Generator	500
Hub	600
Hydraulics	700
Nacelle	800
Over speed system	900
Pitch	1,000
Power factor correction (PFC)	1,100
Tower	1,200
Transmission	1,300
Yaw	1,400
Measurement (sensors etc)	1,500
HV system	1,600
Switchgear	1,700

3.3 Develop Understanding of System under Study

Original equipment manufacturers' (OEM) maintenance and user manuals were used to develop a good understanding of the WT sub-systems in Table 2. This was augmented by expert knowledge of wind farm site operators. Together with maintenance records, a comprehensive system picture was built up. The resultant functional tree for the WT system is omitted for brevity.

3.4 FMECA

A FMECA of model A and B turbines was carried out independently and prior to the work in this chapter. The findings are shown in Table 3 where the risk priority number (RPN) is between 1 (low risk) and 5 (very high). The work presented in this chapter builds on this previous work by providing a higher level of detail in terms of failure modes experienced and quantification of rates of occurrence, as well as economic impact.

4 Main Results

4.1 Data Processing

The data summarised in Table 1 were processed and categorised according to fault type, asset category, turbine model, date stamp and type of maintenance performed.

Table 3 FMECA output-risk ranking

Asset	RPN
Gearbox	4
Transmission	4
Slip ring	2
Hydraulic System	2
Anemometry	2
α-control panel	2
μ-control panel	2
Generator	2
Tower	2
Pitch	2

Downtime information was also available for some failures. The main issue when carrying out the data processing was a lack of standardisation in terms of fault reporting. In some cases information was comprehensive, and in others, highly sparse. Lack of standardisation of component nomenclature was particularly arduous as this meant the data processing could not be automated.

Another issue is the definition of a fault. In some cases faults can be re-set remotely—does this constitute a maintenance entry? Likewise, inspections may be either planned or in reaction to a perceived fault or abnormal operating condition. The maintenance records had to be carefully interpreted in order that mistakes were not made with failure classifications.

Figure 1 shows the *occurrence rate* of maintenance entries by asset category. These include all entries: inspections (planned and reactive), fault investigations, as well as repairs, replacements and retrofits. Because of this, some of the occurrence rates in Fig. 1 are surprisingly large. Gearbox and transmission asset groups in particular are in some cases inflated by early-life inspection regimes to mitigate possible serial defects. Nacelle asset group occurrence rates stem mainly from anemometry. The controller asset group is by far the biggest contributor.

Table 4 shows the *failure rate* per annum, λ, of the most frequently occurring failure modes—that is component failures which require an unscheduled maintenance visit. Remote resets, scheduled inspections, and retrofits are not included. For a more comprehensive analysis, impact of failures should also be included, since the failure modes in Table 4 may not be among the most problematic from a maintenance viewpoint.

4.2 Risk Analysis

The failure rates in Table 4 were calculated for all failure modes in the database. Costs were extracted from three sources. Component costs were obtained from an OEM component spares list. Cranage (hire rates of £ 13k 'call out' and £ 6k per individual day of hire of 500 t crane, suitable of hub heights of less than 85 m) and external labour costs (£ 80/h) were obtained from a wind farm operator. Finally lost revenue was calculated on the basis of a 27 % capacity factor, 2 MW rating (R_{WT}) and production credit of £ 76/MWh. Downtimes were extracted from the database

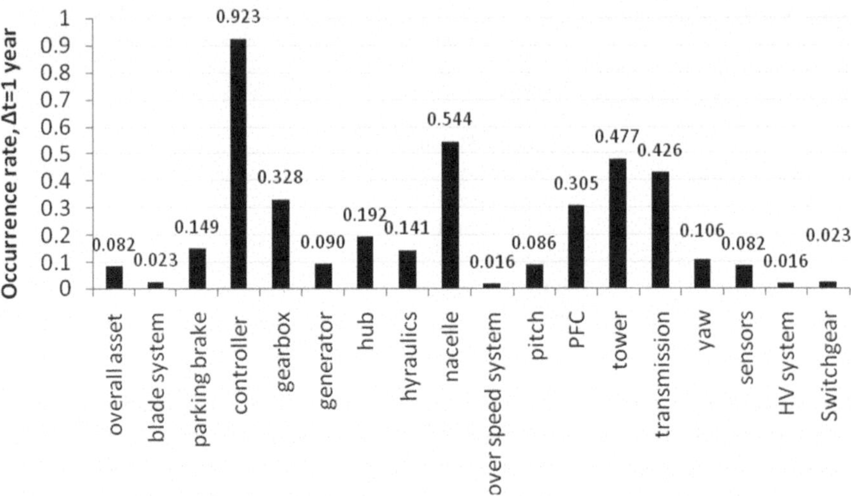

Fig. 1 Annual occurrence rate by asset group

Table 4 Failure modes ranked by failure rate

Failure Mode	*FM#*	λ
μ-panel cooling system pressure error	309.3	0.315
Capacitor bank failure	1,102.1	0.300
μ-panel Grid inverter trip	309.01	0.291
ς-panel yaw converter error	314.6	0.266
α-panel—trip coil fault	301.3	0.125
switching module replacement	1,106	0.116
Wind vane replace	801.9	0.094
ς-panel frequency converter fault	312	0.085
μ-panel cooling system fault	309	0.085
Anemometer cup type replace	801.2	0.082

where possible, and utility experience used to make estimates of downtime where the database information were sparse. The resultant cost class (C_{class}) ranges are summarised in Table 5. Each failure mode was classified according to its cost class. Note that lower range values in Table 5 have been used in the cases presented here.

Table 6 plots the top 10 failure modes by risk level. The quantities—including annual cost of maintenance per wind turbine (CM_{WT}) and annual cost of maintenance per MW (CM_{MW})—were calculated using Eqs. (3, 4, 5 and 6).

$$\text{Risk} = \lambda_{FM\#} \cdot \text{C}_{\text{class}} \tag{3}$$

$$\text{Approx.cost} = \text{Freq}._{\text{FM\#}} \cdot \text{C}_{\text{class}} \tag{4}$$

Table 5 Cost ranges for failures

Class	Cost class (C_{class})£
A	300,000+
B	100,000–300,000
C	50,000–100,000
D	10,000–50,000
E	5,000–10,000
F	1,000–5,000
G	500–1,000
H	100–500

$$CM_{WT} = \frac{\text{total spend for all FM\#}}{WT\Delta t_{total}} \tag{5}$$

$$CM_{MW} = \frac{CM_{WT}}{R_{WT}} \tag{6}$$

These quantitative results align well with qualitative FMECA studies shown earlier in Table 3. The top 3 high risk set of failures are further examined to establish what actions can be taken to mitigate these key failure modes.

Retrofits not included in cost—retrofit campaigns assumed successful

$WT\Delta t_{total}$	255.72
total spend for all FM#	£ 1,012,173
CM_{WT}	£ 3,958
CM_{MW}	£ 1,979

4.3 Corrective Actions

Feeding the results of RCM analysis back to decision making is perhaps the least well-defined area of RCM. Indeed Smith and Hinchcliffe [9] state that "RCM methodology focuses only on what task should be done… [maintenance] intervals are derived from separate analyses".

In practice there are several practical actions a wind farm operator can explore in order to minimise operational risk. For the cracked gearbox failure mode, 95 % of the cost is tied up in component replacement and cranage. Cranage costs could be reduced by good planning of replacement actions, thus avoiding multiple 'call out' cranage rates.

More accurate measures of condition will help operators to plan gearbox replacements in an improved manner. Use of borescopes for improved inspection, and offline oil analysis are two tools which have been used in the aviation industry and can

Table 6 Failure modes ranked by risk

Rank	Failure mode	*FM #*	Frequency	λ	Class	Risk	Approximate cost
1	Cracked gearbox	411.0	1	0.004	A	1,200	£ 300,000
2	μ-panel replace	309.5	8	0.097	D	970	£ 80,000
3	Capacitor bank failure	1,102.1	52	0.300	F	300	£ 52,000
4	HV breaker replace	1,708.0	1	0.004	C	200	£ 50,000
5	High speed shaft replace	403.0	2	0.008	D	80	£ 20,000
6	Wind vane 2 replace	801.9	24	0.094	F	94	£ 24,000
7	μ-panel board replace	309.8	3	0.036	F	36	£ 3,000
8	β-panel—data buffer	302.7	4	0.016	F	16	£ 4,000
9	Yaw gear replace	1,405.1	3	0.012	F	12	£ 3,000
10	Anemometer cup type replace	801.2	21	0.082	H	8.2	£ 2,100

be used on wind turbine gearboxes. The high-risk nature of gearbox failure modes as shown in Table 6 justifies the cost of these outlays (capital cost of borescope is ~£ 25k, and an oil analysis test can be done for as little as £ 10 per sample).

The control panel failures shown in Table 6 have subsequently been traced to moisture sensitivity and handling issues at the supplier end. This failure can be controlled via improved handling and testing of electronic subassemblies.

Capacitor bank failures are problematic in the sense that they can have secondary effects which increase the cost impact. Additionally Table 6 shows that the failure rate is high and should be reduced. This could be achieved either by engaging the OEM to control quality, or sourcing parts from a different supplier. Furthermore the containment of individual capacitors could be improved, in order to stop secondary failures occurring—this would be a feedback to design.

4.4 Proactive Maintenance

The potential benefits of improved operational planning as a result of RCM studies and adoption of more proactive and adaptive maintenance approach such as CbM is illustrated by considering the practical working constraints of a wind farm operator. In particular, the impact of environmental constraints such as wind speeds on maintenance activity, and constraints on daylight working hours (Figs. 2 and 3 respectively) must be considered in any attempt to provide a practical maintenance solution.

Figure 2 provides a general indication of the relative rarity of the suitable weather conditions needed for large maintenance activities such as a nacelle or rotor lift operation at a typical UK onshore site. This points to the need for improved operational planning to secure the services of lifting equipment and key personnel.

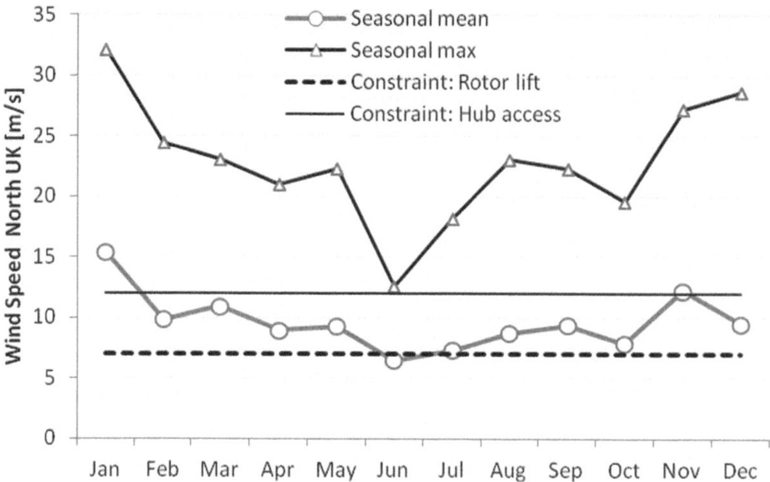

Fig. 2 Seasonal wind farm maintenance constraint: wind speed

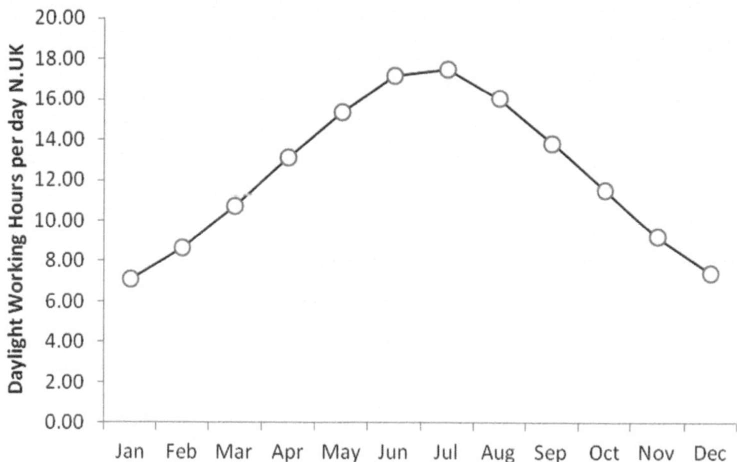

Fig. 3 Seasonal wind farm maintenance constraint: daylight working hours

Daylight working hours will be a major issue for extended periods of work and in particular for offshore operation. Figure 3 shows that in the UK, for unplanned outages occurring in the winter months, repair actions may be seriously constrained by lack of daylight working hours. Again this heightens the requirement for better maintenance planning.

5 Conclusions

The following general observations are made based on the work in this chapter.

5.1 Rigorous Data Entry

Data entry into the maintenance database was of variable quality. In some cases the records were very complete. In other cases, records had omitted information such as date stamps, turbine number, and corrective action taken. Since the intention is to use the maintenance records as a retrospective source of failure information (to calibrate bathtub curves etc.), use of the database will have to become more rigorous.

5.2 Fault Reporting Standardisation

Nomenclature for component failures did not appear to be fully standardised despite the asset group data structure. In many cases failures were placed in the wrong asset category or the component was referred to using shorthand notation of some kind. This made retrospective analysis of failures more difficult. For fleet-wide analysis, nomenclature for components and failure modes should be standardised and used across all sites. These issues are discussed further in [13].

5.3 Calibration of Reliability Curve

One of the long term aims was to examine the relationship between failures and time, to examine if failures were increasing or decreasing over time (perhaps adhering to a 'bathtub' or other characteristic reliability curve [14]). Because of the relatively short data set used, it has not been possible to do this. However it will be possible to track performance in the future using a standardised database. The database can be periodically updated (on a quarterly or bi-annual basis) to chart the reliability performance of individual turbine, site or fleet over time.

5.4 Value of Good Data

The standardisation and data entry issues touch on the wider question of "value of data". Since better information capture involves time and resource, it could be questioned whether or not the extra effort is worthwhile from a cost/benefit viewpoint. It would be beneficial to quantify the value of good data in terms of what it can provide in reduction of operational expenditure. It is essential that future performance improvements enabled by access to good data and initiated by application of RCM tools is measured to provide quantifiable evidence that such a process is economically beneficial [11].

The following specific conclusions are drawn:

- Gearbox failures continue to dominate operational risk in wind turbines.

This brings into sharp focus the need for design robustness, supplier quality control, and in the longer term, cost effective condition monitoring [15]. Factors currently undermining the economic case for online condition monitoring (particularly vibration monitoring) such as false positives and poor fault diagnosis accuracy, will have to be resolved. A large amount of work needs to be done in extracting meaningful information from existing online condition measurements [16]. Until such techniques are developed further, onshore wind farm operators are likely to favour offline oil analysis as discussed in Sect. 4C, and visual inspections carried out as part of scheduled maintenance.

- Rate of occurrence and impact of lower risk failures will increase in the offshore environment.

Table 4 showed that some of the most frequently occurring failures are measurement devices, whose good function is crucial to turbine control and operation. The rate of occurrence of such failures will increase in offshore wind farms due to the more hostile maritime environment. The impact of failure will also substantially increase due to lost production, weather constraints etc. The risk attributed to each failure is therefore very specific to the characteristics of the operating environment. The significance of such failures will be increased offshore—in this case, increased maintenance effort may be impractical. From the design side, functional redundancy could increase to compensate (e.g. multiple anemometers, sensors) however this may not be economically viable. Alternatively a simpler, more robust design could be pursued, though in some cases designing components out of a WT can increase the capital cost and undermine the economic case for adoption of those designs [17]. On the operational side, to aid maintenance planning, a highly refined system of condition monitoring and maintenance management, similar to systems in the aviation sector, will have to be rolled out.

References

1. Ribrant J, Bertling L (2007) Survey of failures in wind power systems with focus on Swedish wind power plants during 1997–2005. IEEE Trans Energy Conver 22(1):167–173
2. Rademakers LWMM, Seebregts AJ, Van den Horn BA, Jehee JNT, Blok BM (1993) Methodology for probabilistic safety assessment of wind turbines, ECN-C-93–010
3. Michos D, Dialynas E, Vionis P (2002) Reliability and safety assessment of wind turbines control and protection systems. Wind Eng 26(6):359–369
4. Arabian-Hoseynabadi H, Oraee H, Tavner PJ (2010) Failure modes and effects analysis (FMEA) for wind turbines. Electr Power Energy Syst 32:817–824

5. Andrawus JA, Watson J, Kishk M, Adam A (2006) The selection of a suitable maintenance strategy for wind turbines. Wind Eng 30(6):471–486
6. McMillan D, Ault GW (2008) Quantification of condition monitoring benefit for onshore wind turbines: sensitivity to operational parameters. IET Renew Power Gener 2(1):60–72
7. Besnard F, Bertling L (2010) An approach for condition-based maintenance optimization applied to wind turbine blades. IEEE Trans Sustain Energy 1(2):77–83
8. Byon E, Ntaimo L, Ding Y (2010) Optimal maintenance strategies for wind turbine systems under stochastic weather conditions. IEEE Trans Reliab 59:393–404
9. Smith AM, Hinchcliffe GR (2004) RCM—gateway to world class maintenance. Elsevier
10. Moubray J (1992) Reliability-centered maintenance. Industrial Press Inc, New York
11. Guevara Carazas FJ, Martha de Souza GF (2009) Availability analysis of gas turbines used in power plants. Int J of Thermodyn 12(1):28–37
12. MAXIMO. http://www-01.ibm.com/software/tivoli/products/maximo-asset-mgmt/
13. Fischer K, Besnard F, Bertling L (2011) A limited-scope reliability-centred maintenance analysis of wind turbines. In: EWEA, Brussels, Belgium, 14–17 March 2011
14. Hjorth U (1980) A reliability distribution with increasing, decreasing, constant and bathtub-shaped failure rates. Technometrics 22(1):99–107
15. Yang W, Tavner PJ, Crabtree CJ, Wilkinson M (2010) Cost-effective condition monitoring for wind turbines. IEEE Trans Ind Electron 57(1):263–271
16. Gray CS, Watson SJ (2010) Physics of failure approach to wind turbine condition based maintenance. Wind Energy 13(5):395–405
17. McMillan D, Ault GW (2010) Techno-economic comparison of operational aspects for direct drive and gearbox-driven wind turbines. IEEE Trans Energy Convers 25(1):191–198

Cable Segment Replacement Optimization

Patrik Hilber

1 Introduction

Today many distribution system operators (DSOs) face ageing equipment. Even though most of the equipment is written off; the equipment represents significant values, at least in terms of reinvestment costs. Therefore, it becomes crucial for the DSOs to replace the equipment in a cost efficient way, i.e., to replace the most critical parts first and to utilize "healthy" equipment as far as economically beneficial. To address these kinds of problems diagnostic methods are developed. Examples of such methods are the underground power cable diagnostic methods presented in [2]–[4]. These methods not only give us indications on the status of cables, but also an indication of the status along the cable. For the case of new cables, temperature sensors can be obtained at a relatively small cost increase. With temperature sensors along the cable, a good estimate on the cable condition can also be obtained. The major problem with temperature sensors is that the diagnostic equipment at the terminal is expensive. This is however likely to change, with measurement apparatus becoming more affordable over time. The overall conclusion is that it becomes important to use data that can be obtained.

The concept presented here is to utilize the information from cable diagnostics and combine it with information on replacement costs and component reliability importance [5]. These factors are all considered under a budget constraint.

The major costs considered in the model are replacement costs per segment and a startup cost for replacements. The segment cost is based on a cost per meter of cable replaced, including cable cost, digging, and other work related to the length of the segment and the environment it goes through. The studied optimization problem becomes opportunistic because of the applied starting cost. When a replacement is initialized there are significant costs related to the startup, e.g., to bring machinery to place and administration work to be performed. This is modeled as a cost that is

P. Hilber (✉)
Electromagnetic Engineering, KTH, Royal Institute of Technology, Stockholm, Sweden
e-mail: hilber@kth.se

R. Karki et al. (eds.), *Reliability Modeling and Analysis of Smart Power Systems*, Reliable and Sustainable Electric Power and Energy Systems Management,
DOI 10.1007/978-81-322-1798-5_13, © Springer India 2014

added once for every series of connected segments, i.e., it becomes cheaper to replace cable segments in contact with the segment where you are currently working, no startup cost for additional segments.

2 Optimization Model

The aim of the model presented below is to minimize the customer disutility by means of replacing cable segments under the constraint of a budget. Every cable has its own significance for the network, i.e., a component reliability importance that corresponds to the customer interruption cost (disutility) in case of a failure of the cable. Every segment has its own failure rate and based on the minimization of disutility and the assumptions below the model is proposed.

Assumptions:

- The cables are dividable into segments, in the example study of this chapter the cable segments are 10 m, this is not a limitation in the model, and the segments may have variable length.
- There is a startup cost for replacement of a cable segment; this startup cost is only invoked once for a series of connected segments being replaced.
- Every cable segment has its own replacement cost. This value depends on the length of the cable segment, type of cable, and environment that the cable runs through, for example if the segment goes along a road, in a forest area or under a highway.
- Every segment has its own failure rate or other relative degradation rating.
- Every cable has its own importance for its network; this value may be expressed in terms of expected customer interruption cost in case of failure.
- The component reliability importance of involved cables are independent.
- There is a cable replacement budget.
- A replaced cable segments failure rate is assumed to become zero (i.e., constant/random failures are removed from the calculations).

Problem formulation:

$$\min \sum_{i=1}^{N} I_i \sum_{j=1}^{L} s_{ij}(1 - x_{ij}) \quad (1)$$

Subject to:

$$\sum_{i=1}^{N} \sum_{j=1}^{L} (cy_{ij} + P_{ij}x_{ij}) \leq B \quad \forall i \in 1 \ldots N, j \in 1 \ldots L \quad (2)$$

$$y_{ij} \geq x_{ij} - x_{i(j+1)} \quad \forall i \in 1 \ldots N, j \in 1 \ldots L \quad (3)$$

$$x_{ij} = 0 \quad \forall i \in 1 \dots N, j \in L+1 \tag{4}$$

$$x_{ij}, y_{ij} \in \{0,1\} \quad \forall i, j \in \aleph \tag{5}$$

Where

N Number of cables
L Maximum number of segments in any cable
i Cable number
j Segment number of cable (part of cable)
I_i Component reliability importance of cable i
C Startup cost for replacement
s_{ij} Status (failure rate) of segment ij
P_{ij} Replacement cost of segment ij
B Budget constraint
x_{ij} Binary variable, '1' indicates replacement, main output from the optimization
y_{ij} Binary variable, '1' once for every startup

Equation outline:

1. Objective function, calculates the total expected customer inconvenience (customer interruption cost) due to cable failures.
2. Cable replacement budget constraint, startup cost plus segment replacement cost.
3. Constraint that states that if x goes from 1 to 0 in the transition from j to $j+1$, y_{ij} becomes 1 and thereby records a new startup cost. Hence (Eq. 3) helps us capture the number of startup costs.
4. Ensures that all startup costs are recorded.
5. States that x and y are binary variables.

Equation (1) implicitly states that a replaced cable segment obtains a failure rate reduced to zero. Hence, a replaced cable segment does not contribute to the customer interruption cost. This formulation is motivated by that only condition related failures are studied in the model. Equation (4) is a necessary border constraint in order to be able to perform the calculation (Eq. (3)) for every segment. The reader may note that j in Eq. (3) and Eq. (4) passes the maximum number of segments in any cable.

The component reliability importance, I, represents the power system in the problem. Component reliability importance indices have been developed specifically for power systems [5] and it is suggested to use an index that indicates the systems sensitivity towards changes in the component failure rate, typically expressed as either a customer cost function or in a combination of other customer parameters such as number of expected interruptions and energy not served. The status, s, of the component is based on diagnostics and possibly environment. In the proposed model it is assumed that failure rates are known or that a relative ranking between cable segments exists (see section "Discussion and future work").

Table 1 Cable reliability importance

Cable	1	2	3	4	5
Importance [€/f]	200	500	1,000	1,000	3,000

Table 2 Failure rates per segment

Cable segments	1	2	3	4	5
1	1	16	1	1	1
2	2	2	2	2	2
3	4	4	4	4	4
4	8	8	8	8	8
5	16	16	16	16	16
6	8	8	8	8	16
7	4	4	4	4	8
8	2	2	2	2	4
9	1	1	1	1	2
10	16	0	0	16	1
11	1	0	0	1	0

Segment status (10^{-3} f/year), where 0 is failure free and "16" is the worst degraded status. The values might be interpreted as failure rates for every segment

3 Case, Demonstration Problem

A brief demonstration problem is analyzed in order to display the properties of the method. The cable population in this chapter is an example population; costs used are based on costs for an 11 kV distribution system. The example population consists of 5 cables of up to 11 segments each.

3.1 Cable Reliability Importance

Every cable has its own reliability importance for the network (or possibly their respective network), presented in Table 1. The reliability importance is expressed in terms of cost of a failure (€/f), see for example [5] and [6] for the calculation of component reliability importance indices. The reliability importance of the cables are assumed to be independent of each other. It is noteworthy that the relative difference between the values in Table 1 are low compared to the values presented in [6].

3.2 Cable Status

In Table 2 the status of every cable segment is presented. The values are chosen in order to demonstrate how the optimization works and are in the current setup only interesting in terms of relative values with respect to each other. A value of "1" might be considered as the first instance of degradation and "16" as the last (worst). In the current setup the values in Table 2 are considered to be failure rates [f/year].

Table 3 Segment replacement cost

Cable segment	1	2	3	4	5
1	400	400	400	400	1,200
2	800	400	400	400	1,200
3	1,200	400	400	400	1,200
4	400	400	400	400	1,200
5	400	400	400	400	1,200
6	400	400	400	400	400
7	400	400	400	400	400
8	400	400	400	400	400
9	400	400	400	400	400
10	400	400	400	400	400
11	400	400	400	400	400

Cost of cable segment replacement in Euro, without startup cost

In a more realistic case it might not be possible to establish true failure rates but "only" relationships between the status of segments'; hence, we keep the status (Sij) open for not only failure rates but also other estimates of the status. (Nevertheless, the model becomes theoretically more interesting when Sij denotes the failure rate).

3.3 *Costs*

The budget is set to 15,599 €. Table 3 presents costs in Euro for cable segment replacement, without startup cost. The values are based on segments of 10 m length and a cost of 40,000 €/km as a basic replacement cost. The higher values, i.e., >400 €, represent environment such as infrastructure that has to be rebuilt, roads for example. In previous studies, costs for replacement of 12 kV cables stretch from 20,000 €/km in rural areas to above 100,000 €/km in urban areas, see [7]. These values indicate the significance of the environment that the cable goes through for its replacement cost.

The startup cost for replacement of a cable segment or a series of connected cable segments is in the model set to 2,000 €, this cost corresponds to the procedure of bringing workforce and equipment to the start of cable replacement and paperwork that has to precede and succeed a replacement.

3.4 *Optimization*

The resulting integer property optimization problem is implemented in the AMPL-language™. And the problem solved with the CPLEX™ solver called form within AMPL.

To solve the current problem CPLEX required 38 MIP simplex iterations and zero branch and bound operations. This may indicate that the problem is relatively easy to solve for the solver since it never goes into the branch and bound algorithm. The problem is; however, relatively small and might become more complicated for

Table 4 Results, "x"

Cable segment	1	2	3	4	5
0	0	0	0	0	0
2	0	0	0	0	0
3	0	0	0	0	1
4	0	0	1	1	1
5	0	0	1	1	1
6	0	0	1	1	1
7	0	0	0	1	1
8	0	0	0	1	1
9	0	0	0	1	1
10	0	0	0	1	0
11	0	0	0	0	0

the solver with more segments to analyze. Basically the problem becomes an extension of the knapsack problem.

3.5 *Results*

The results from the optimization are presented in Table 4, where "1" denotes a replacement. The three most important cables are the ones that the solver (or rather the solution to the problem) suggests replacements for. It can be seen that the startup cost is only invoked three times. One example where the startup cost affects the solution is when segment 9 of cable 4 is replaced while segment 5 for cable 2 is not, even though the replaced segment "only" removes 10 from the objective contra the 80 that would have been removed by segment 5 in cable 2.

The slack variable for Eq. (2) becomes 399 for the studied case. The budget constraint was deliberately chosen in order to test the solver.

One more interesting thing to note is that replacements are suggested for segments in cable 3 and not for segment 1 in cable 5 which has a high failure rate. Even though cable 5 is three times more important, this is precisely cancelled out by the higher replacement cost of segment 1 on cable 5. This is due to that segment 1 and 2 in cable 5 are more expensive to replace.

4 Discussion and Future Work

The main merit of this chapter is the combination of diagnostics, economics, and power system (customers and topology), in one model. The model could be further developed as indicated below, but focus should rather be placed on the connection of diagnostics to status of the cable. Methods exist for determining the status, for example by location of partial discharges and determining location of joints and cuts in the screening [4]. These indications on degradation or damage to the cable can be used to rank individual segments by an expert. One approach to establish the failure rates for the segments might be to use an estimate on the total failure rate for the whole population, based on statistics from similar components, data on the population's general status and local cable failures. Then use the data from diagnostics to distribute the failure rate over the segments, probably with respect to the environment. This approach obviously needs development in collaboration with diagnostic experts. Another, related approach is to only work with relative status of the involved segments. For example, by specifying critical partial discharge levels and put them in relation to criticality of joints and cuts.

Expert knowledge is moreover needed for the system analysis, in order to obtain component reliability importance and establishing replacement costs. Finally, economy provides a framework for the whole model and sets the upper limit of investments. The concept here is that it should be sufficient that diagnostics only provides relative values for the cable status, and then by combining this information with system knowledge and economics, a decision closer to the optimal asset performance can be made.

Improvements to the model could be to model time and costs more detailed. By introducing a time aspect, cable degradation models could be used. The time aspect present a more difficult optimization problem, e.g., determining when in time replacements should occur and uncertain "replacement budgets" from year to year. Caution is important when considering implementation of time into the model, since a degradation model requires more data. Data that is hard to establish values for. Improvements on the cost side are more straightforward, an improved cost model would probably enhance the realism of the model, e.g., by inclusion of costs for corrective maintenance in case of cable failure. Additional rules that could be interesting to implement is a limit to the number of startups and that if more than a certain percentage of a cable is suggested for replacement, the whole cable should be replaced. Such restrictions will; however, produce more costly solutions. An alternative development is to penalize new joints in the objective function with a cost (this is partly covered with the startup cost, C).

The chosen segment length in the example studied above is something that might be altered significantly in a case study based on more real data. This would probably result in longer segments, where practical issues demands for replacement (or not) of longer stretches of cable.

5 Conclusions

This chapter presents a straightforward framework for identifying interesting cable segments for replacement in a cable population. In this, utilizing results from diagnostic methods for cables, and incorporating network structure and economy. With further studies on how to use the information gained from the diagnostic measurements this framework provides a rational base in the search for the optimal replacement of cable segments.

References

1. Hilber P (2012) Cable segment replacement optimization. Proceedings of the 12th probabilistic methods applied to power systems (PMAPS), Istanbul, Turkey
2. Steiner JP, Reynolds PH (1992) Estimating the location of partial discharges in cables. IEEE Trans Electr Insul 27(1):44–59
3. Stone GC (2005) Partial discharge diagnostics and electrical equipment insulation condition assessment. IEEE Trans Dielectr Electr Insul 12(5):891–904
4. Dubickas V (2009) Development of on-line diagnostic methods for medium voltage XLPE power cables. PhD thesis, ISBN 978-91-7415-220-3, KTH, Stockholm
5. Hilber P (2008) Maintenance optimization for power distribution systems. PhD Thesis, KTH, Stockholm, Sweden, ISBN 978-91-628-7464–3
6. Hilber P, Bertling L (2004) Monetary importance of component reliability in electrical networks for maintenance optimization. Proceedings of the 8th International Conference on Probabilistic Methods Applied to Power Systems, PMAPS 2004, Ames, USA
7. Swedenergy (2008) EBR—cost catalog local networks 0.4–24 kV and opto networks, in Swedish. KLG 1:08, March 2008

The manufacturer's authorised representative in the EU is Springer Nature Customer Service Centre GmbH, Europaplatz 3, 69115 Heidelberg, Germany. If you have any concerns regarding our products, please contact ProductSafety@springernature.com

Printed and bound by CPI Group (UK) Ltd, Croydon, CR0 4YY
17/07/2026
02170245-0003